DEVELOPING A ROADMAP FOR THE COMMERCIAL DEPLOYMENT OF NUCLEAR HYDROGEN PRODUCTION

The following States are Members of the International Atomic Energy Agency:

AFGHANISTAN
ALBANIA
ALGERIA
ANGOLA
ANTIGUA AND BARBUDA
ARGENTINA
ARMENIA
AUSTRALIA
AUSTRIA
AZERBAIJAN
BAHAMAS, THE
BAHRAIN
BANGLADESH
BARBADOS
BELARUS
BELGIUM
BELIZE
BENIN
BOLIVIA, PLURINATIONAL STATE OF
BOSNIA AND HERZEGOVINA
BOTSWANA
BRAZIL
BRUNEI DARUSSALAM
BULGARIA
BURKINA FASO
BURUNDI
CABO VERDE
CAMBODIA
CAMEROON
CANADA
CENTRAL AFRICAN REPUBLIC
CHAD
CHILE
CHINA
COLOMBIA
COMOROS
CONGO
COOK ISLANDS
COSTA RICA
CÔTE D'IVOIRE
CROATIA
CUBA
CYPRUS
CZECH REPUBLIC
DEMOCRATIC REPUBLIC OF THE CONGO
DENMARK
DJIBOUTI
DOMINICA
DOMINICAN REPUBLIC
ECUADOR
EGYPT
EL SALVADOR
ERITREA
ESTONIA
ESWATINI
ETHIOPIA
FIJI
FINLAND
FRANCE
GABON
GAMBIA, THE
GEORGIA
GERMANY
GHANA
GREECE
GRENADA
GUATEMALA
GUINEA
GUYANA
HAITI
HOLY SEE
HONDURAS
HUNGARY
ICELAND
INDIA
INDONESIA
IRAN, ISLAMIC REPUBLIC OF
IRAQ
IRELAND
ISRAEL
ITALY
JAMAICA
JAPAN
JORDAN
KAZAKHSTAN
KENYA
KOREA, REPUBLIC OF
KUWAIT
KYRGYZSTAN
LAO PEOPLE'S DEMOCRATIC REPUBLIC
LATVIA
LEBANON
LESOTHO
LIBERIA
LIBYA
LIECHTENSTEIN
LITHUANIA
LUXEMBOURG
MADAGASCAR
MALAWI
MALAYSIA
MALI
MALTA
MARSHALL ISLANDS
MAURITANIA
MAURITIUS
MEXICO
MONACO
MONGOLIA
MONTENEGRO
MOROCCO
MOZAMBIQUE
MYANMAR
NAMIBIA
NEPAL
NETHERLANDS, KINGDOM OF THE
NEW ZEALAND
NICARAGUA
NIGER
NIGERIA
NORTH MACEDONIA
NORWAY
OMAN
PAKISTAN
PALAU
PANAMA
PAPUA NEW GUINEA
PARAGUAY
PERU
PHILIPPINES
POLAND
PORTUGAL
QATAR
REPUBLIC OF MOLDOVA
ROMANIA
RUSSIAN FEDERATION
RWANDA
SAINT KITTS AND NEVIS
SAINT LUCIA
SAINT VINCENT AND THE GRENADINES
SAMOA
SAN MARINO
SAUDI ARABIA
SENEGAL
SERBIA
SEYCHELLES
SIERRA LEONE
SINGAPORE
SLOVAKIA
SLOVENIA
SOMALIA
SOUTH AFRICA
SPAIN
SRI LANKA
SUDAN
SWEDEN
SWITZERLAND
SYRIAN ARAB REPUBLIC
TAJIKISTAN
THAILAND
TOGO
TONGA
TRINIDAD AND TOBAGO
TUNISIA
TÜRKİYE
TURKMENISTAN
UGANDA
UKRAINE
UNITED ARAB EMIRATES
UNITED KINGDOM OF GREAT BRITAIN AND NORTHERN IRELAND
UNITED REPUBLIC OF TANZANIA
UNITED STATES OF AMERICA
URUGUAY
UZBEKISTAN
VANUATU
VENEZUELA, BOLIVARIAN REPUBLIC OF
VIET NAM
YEMEN
ZAMBIA
ZIMBABWE

The Agency's Statute was approved on 23 October 1956 by the Conference on the Statute of the IAEA held at United Nations Headquarters, New York; it entered into force on 29 July 1957. The Headquarters of the Agency are situated in Vienna. Its principal objective is "to accelerate and enlarge the contribution of atomic energy to peace, health and prosperity throughout the world".

IAEA Nuclear Energy Series No. NR-G-4.1

DEVELOPING A ROADMAP FOR THE COMMERCIAL DEPLOYMENT OF NUCLEAR HYDROGEN PRODUCTION

INTERNATIONAL ATOMIC ENERGY AGENCY
VIENNA, 2025

COPYRIGHT NOTICE

Publishing Section
International Atomic Energy Agency
Vienna International Centre
PO Box 100
1400 Vienna, Austria
tel.: +43 1 2600 22529 or 22530
email: sales.publications@iaea.org
www.iaea.org/publications

Printed by the IAEA in Austria
September 2025
STI/PUB/2096

https://doi.org/10.61092/iaea.qp5y-yn4t

IAEA Library Cataloguing in Publication Data

Names: International Atomic Energy Agency.
Title: Developing a roadmap for the commercial deployment of nuclear hydrogen production / International Atomic Energy Agency.
Description: Vienna : International Atomic Energy Agency, 2025. | Series: IAEA nuclear energy series, ISSN 1995-7807 ; no. NR-G-4.1 | Includes bibliographical references.
Identifiers: IAEAL 24-01706 | ISBN 978-92-0-126424-4 (paperback : alk. paper) | ISBN 978-92-0-126524-1 (pdf) | ISBN 978-92-0-126624-8 (epub)
Subjects: LCSH: Hydrogen as fuel. | Hydrogen as fuel — Research. | Hydrogen industry — Economic aspects.
Classification: UDC 662.769.2 | STI/PUB/2096

FOREWORD

The IAEA's statutory role is to "seek to accelerate and enlarge the contribution of atomic energy to peace, health and prosperity throughout the world". Among other functions, the IAEA is authorized to "foster the exchange of scientific and technical information on peaceful uses of atomic energy". One way this is achieved is through a range of technical publications including the IAEA Nuclear Energy Series.

The IAEA Nuclear Energy Series comprises publications designed to further the use of nuclear technologies in support of sustainable development, to advance nuclear science and technology, catalyse innovation and build capacity to support the existing and expanded use of nuclear power and nuclear science applications. The publications include information covering all policy, technological and management aspects of the definition and implementation of activities involving the peaceful use of nuclear technology. While the guidance provided in IAEA Nuclear Energy Series publications does not constitute Member States' consensus, it has undergone internal peer review and been made available to Member States for comment prior to publication.

The IAEA safety standards establish fundamental principles, requirements and recommendations to ensure nuclear safety and serve as a global reference for protecting people and the environment from harmful effects of ionizing radiation.

When IAEA Nuclear Energy Series publications address safety, it is ensured that the IAEA safety standards are referred to as the current boundary conditions for the application of nuclear technology.

More than 95% of the hydrogen used today is produced from fossil resources (primarily gas, but also oil and coal), with adverse effects such as carbon dioxide emissions and resource depletion. Nuclear energy as a low carbon energy source has the potential to replace fossil fuels for supplying the forecasted increasing demand for hydrogen. There are several ongoing and planned demonstration projects worldwide to produce hydrogen using operational nuclear power plants, as well as developments related to advanced reactor technologies, including high temperature reactors. Various hydrogen generation options are under consideration worldwide for coupling with nuclear reactors, including conventional electrolysis, high temperature steam electrolysis and thermochemical cycles. Steam methane reforming is also being considered in view of lowering fossil fuel consumption by using nuclear reactors to provide the necessary energy input for the process.

The safety and licensing of nuclear hydrogen production projects is an important issue to be explicitly addressed during all phases of the licensing, from siting to design to safety assessment. Hydrogen production may represent a hazard to the nuclear island in terms of the potential for explosive gases, fire and/or projectiles, thus special emphasis is given to those cogeneration facilities that are coupled with the nuclear island and to the safety analysis to be developed in the licence application after a detailed safety assessment. Similarly, plant performance, reliability and sustained economic viability will have to be addressed, since the introduction of hydrogen production, especially in the context of currently operating plants, could necessitate plant modifications, procedure revisions, new or revised maintenance practices and additional training. These changes have to be fully understood from perspectives of technical plant integration and of the development of a commercial business case and cost recovery within the plant's remaining operational lifetime. Developing a roadmap for commercial deployment of nuclear hydrogen production would benefit interested Member States, providing a useful management tool for evaluating, planning and strategizing the development of nuclear hydrogen projects.

The IAEA officers responsible for this publication were A. Constantin and F. Ganda of the Division of Nuclear Power and A. Van Heek of the Division of Planning, Information and Knowledge Management.

CONTENTS

1. INTRODUCTION

1.1. BACKGROUND

Achieving clean energy, decarbonization and net zero is a generational challenge that requires a comprehensive and holistic view of the global energy transition. In this context, hydrogen has been acknowledged as a key enabler of decarbonization for its versatility, for its use as feedstock for industrial processes, and for its potential as a clean energy vector and synthetic fuel [1–3]. Currently, various Member States are seeking an improved understanding of how to transition to a hydrogen based energy infrastructure. To truly unlock the potential of hydrogen, its production, distribution, transport and use have to be clean, but also cost-competitive, making use of renewable sources and other low carbon electricity and heat sources, such as nuclear power.

Large scale, low carbon hydrogen production can benefit from the unique attributes of existing nuclear reactors, and the fleet of new nuclear reactors worldwide. Nuclear reactors provide reliability as well as electricity and heat that can be coupled with existing and upcoming technologies for clean hydrogen generation.

Currently the technologies for producing hydrogen at a large scale using clean energy and electricity rely heavily on water electrolysis, but in the future thermochemical cycles might also be considered. Water electrolysis is a mature and readily deployable technology, but deployment of electrolysers at the scales required for global deep decarbonization would be challenging [4]. Large scale production of hydrogen also requires its interconnection with other chemical and metallurgical processes where hydrogen is or can be used in large amounts, such as in the production of ammonia, steel, petrochemicals and methanol. In doing so, hydrogen's utility increases manyfold as it becomes an enabler of industrial growth and decarbonization and not just a commodity product.

Employment of large scale hydrogen production facilities has to consider the relevant safety aspects, including potential leaks, material embrittlement and possible fire hazards and explosions. However, it is important to note that hydrogen safety culture is mature and uses well established and peer reviewed codes and standards. When considering nuclear hydrogen production — that is, production of hydrogen using nuclear energy in the form of electricity, steam or heat — demonstration projects that couple a nuclear power plant with a hydrogen production facility are of crucial importance. These projects would analyse the feasibility of and the business opportunities and potential risks associated with larger scale projects. As in the case of hydrogen produced by renewables, the opportunity for deployment of nuclear hydrogen projects is driven by market developments to address decarbonization policies.

Wide commercial deployment of nuclear hydrogen can be achieved by addressing the following key steps:

(1) Mapping demand and potential hydrogen end users at national, regional and international levels. This would also involve assessment of the potential for hydrogen import and export. Those areas with industrial needs for large scale hydrogen and hydrogen derived products are to be identified at this stage.
(2) Mapping the available and emerging nuclear technologies to reflect the nuclear assets that can be capitalized for nuclear hydrogen production. At this stage, it should be determined which nuclear assets are located or could be built in the near future in the vicinity of industrial regions or clusters with substantial hydrogen demand.
(3) Mapping currently available and emerging hydrogen production technologies to identify the optimal coupling with the nuclear assets identified in the previous step, to deploy nuclear hydrogen projects.
(4) Identifying and addressing the specificities, constraints, challenges and gaps for all dimensions involved for each nuclear hydrogen project. These include technology, business case, safety, licensing, human resource development, supply chain, stakeholder engagement and public acceptance.
(5) Establishing the optimal coupling of nuclear reactors with hydrogen production facilities.
(6) Establishing a safety and licensing framework for each nuclear hydrogen project through efficient interaction between the regulators of both the nuclear power plant and the hydrogen production plant and the other stakeholders involved.
(7) Developing a comprehensive business case for the projects that clearly determines the strategic context, the economic analysis, the commercial approach, the financial case and the management approach. This will establish the grounds for investments in the project and will be used in the interaction with relevant stakeholders, including decision makers and investors.
(8) Ensuring efficient stakeholder engagement and building solid partnerships that will support the deployment of the nuclear hydrogen projects.
(9) Securing the project funding by establishing contracts with the relevant stakeholders.
(10) Deploying demonstration projects to prove the synergistic, reliable and cost-effective coupling of hydrogen production facilities with nuclear reactors.

(11) Expanding the demonstration projects to large scale projects for the various uses of hydrogen, such as the production of ammonia, steel and synthetic fuels, and for energy storage and distribution, while maintaining market sector competitiveness by having high capacity factors, high reliability and energy independence secured by coupling with nuclear reactors. The deployment of nuclear hydrogen projects may be strengthened and facilitated by the introduction of carbon credits, carbon incentives and other means that will accelerate the adoption of clean energy technologies and ensure a level playing field for these technologies.

(12) Considering other opportunities for using the hydrogen produced, such as exports.

(13) Developing and maintaining lasting, synergistic alliances and partnerships with various stakeholders to support the nuclear hydrogen projects and their further expansion.

1.2. OBJECTIVE

The objective of this publication is to provide a useful management tool for evaluating, planning and strategizing the development of nuclear hydrogen projects by interested Member States, considering the technology readiness level (TRL) and time to market of different technologies. It is intended for experts in the field of the hydrogen economy and in decarbonization using hydrogen, as well as interested parties in nuclear hydrogen projects, including executives and strategic decision makers within utilities that are currently operating nuclear power plants, government policy makers and hydrogen end users. Guidance and recommendations provided here in relation to identified good practices represent expert opinion but are not made on the basis of a consensus of all Member States.

1.3. SCOPE

This publication covers the role of the hydrogen economy in global decarbonization efforts, as well as the potential for nuclear energy to support the production of hydrogen to meet the predicted increased demand. Many countries have developed or are developing national roadmaps for hydrogen generation and use. However just a few include the option of nuclear hydrogen production. The publication outlines the status of technologies available to support nuclear hydrogen projects and discusses challenges associated with the process of coupling nuclear energy and hydrogen production technologies. Several dimensions — economics, technology, safety and licensing, as well as stakeholder engagement — are analysed, in order to establish a roadmap for the commercial production of nuclear hydrogen.

1.4. STRUCTURE

Following this introduction, Section 2 outlines the general context of the hydrogen economy and the net zero goals, including current challenges for the global energy system; decarbonization opportunities provided by low carbon hydrogen; and technical, economic and regulatory challenges to deploying hydrogen at scale.

Section 3 is dedicated to the commercial deployment aspects of nuclear hydrogen projects. It presents the commercial drivers for nuclear hydrogen deployment and considers aspects such as the global hydrogen market, the potential of nuclear energy to support the hydrogen market, cost considerations of hydrogen production, the competitiveness of nuclear hydrogen production, the economic perspectives of collocation and coupling of a hydrogen production plant with a nuclear power plant, possible business models, the financing of nuclear hydrogen projects, and barriers and enablers for the nuclear industry in the hydrogen market.

Section 4 presents the technologies for nuclear hydrogen production. It looks at the existing hydrogen production processes suitable for coupling with current and advanced nuclear reactors, the status of integration with nuclear reactors and the associated challenges, and underlines the nuclear hydrogen demonstration projects that are currently planned.

Section 5 describes the safety and regulatory aspects of a nuclear hydrogen project. Special emphasis is given to cogeneration facilities that are coupled with the nuclear island and to the safety analysis to be developed in the licence application after a detailed safety assessment.

Section 6 details stakeholder engagement in a nuclear hydrogen project, considering the key organizations involved in such a project (e.g. operators of nuclear power plants, technology developers, regulatory bodies, government) and the associated uncertainties, major barriers, opportunities, challenges and acceleration actions.

Section 7 provides a generic roadmap for commercial deployment of nuclear hydrogen projects. It includes considerations for the planning and deployment of nuclear hydrogen production projects and describes a generic phase-wise workstream based approach.

The Annex provides examples of national roadmaps to commercial deployment of nuclear hydrogen production from three Member States: Canada, India and Japan.

2. THE HYDROGEN ECONOMY AND THE NET ZERO GOALS

Global warming and human induced or anthropogenic climate change are direct consequences of the emissions of carbon dioxide (CO_2) and other greenhouse gases (GHGs) such as methane, sulphur and nitrogen oxides from the use of fossil fuels [5]. Climate change manifests itself in multiple ways, most notably in weather related events, including changes in rain and snowfall patterns, extreme heat and cold wave events; loss of glacial and polar ice mass; increased occurrence of natural disasters including floods, droughts and forest fires; degradation of soil health; increasing desertification; decreases in agricultural output; and mass extinction of species leading to massive biodiversity loss [6].

2.1. GLOBAL ENERGY ISSUES, CLIMATE CHANGE AND THE NEED FOR LOW CARBON ENERGY TRANSITION

This section introduces the global energy issues in relation to climate change goals and the urgent need for the transition to low carbon energy.

2.1.1. Challenges for the global energy system

The recently issued series of documents making up the Sixth Assessment Report of the Intergovernmental Panel on Climate Change (IPCC) [7, 8] provide undeniable evidence that human actions have led to climate change. They indicate that under the current business as usual growth/development trajectories that most countries are still pursuing, the remaining carbon budget of the Earth's atmosphere (which is about 420 Gt C according to the latest estimates) to limit global temperature rise to 1.5°C above pre-industrial levels is very likely to be exceeded within this decade [7]. A warming of 1.2°C was already confirmed by the IPCC reports in 2021 and 2022, supported by numerous measurements across the world [9, 10]. It is widely accepted that in order to achieve temperature stabilization well below 2°C, the CO_2 level cannot exceed an average value of 450 ppm. This level provides a 67% chance of avoiding drastic temperature rise and its consequences [7]. Thus, an urgent reduction of CO_2 emissions at the global scale is one of the most important immediate actions for climate change mitigation, adaptation and sustainable, inclusive development.

Alongside the challenges of environmental degradation, dependence on fossil fuels has also created potential energy security and sovereignty issues for countries that are heavily dependent on energy imports. Geopolitical issues leading to energy shortages and escalation of fossil fuel costs tend to worsen energy poverty concerns. Thus, energy availability, affordability and reliability, and sustainability of supply have become key drivers of the shift away from fossil fuels towards a low carbon energy system.

The energy sector (including the use of fossil fuels for electric power production; as transportation fuels and thermal energy sources; and for heating, cooling, ventilation and lighting services in the building sector) is the largest contributor to GHG emissions globally (with about 73% of current total global emissions attributed directly to fossil fuel usage) and thus will need to make the greatest transition efforts towards climate action [9]. Similarly, the process and metallurgical industries using fossil based energy materials as feedstock or raw material (e.g. coal for reduction of iron ore in the steel industry, natural gas for producing hydrogen in the ammonia industry, naphtha for chemicals/petrochemicals production) or for utility services (e.g. heating industrial furnaces with natural gas, raising process steam via coal or furnace oil fired boilers) have to play a significant role in taking climate action through major shifts in their material and energy consumption patterns. Ensuring energy availability to support human welfare while achieving climate action targets (both mitigation and adaptation) will therefore require judicious deployment of all low carbon energy forms, including renewables, hydropower and nuclear power and new energy vectors in place of fossil fuels to meet sector specific decarbonization targets.

2.1.2. The role of nuclear power in net zero pathways

The International Energy Agency (IEA) has performed assessments of several energy transition pathways. It identified a pathway to achieve net zero carbon emissions by 2050 [10] in which global deployment of renewable energy systems would have to reach at least 630 GW of solar photovoltaic (PV) energy and 390 GW of wind energy every year until 2030. In the IEA Announced Pledges Scenario for reaching global net zero by 2050, 70% of the electricity generation would be supplied by renewables and nuclear power would have a share of about 10% [10]. In that scenario, new nuclear capacity additions of about 4.5 GW per year from 2021 to 2035 would be required at the global level, along with lifetime extensions and long term operation of many existing reactors. The International Renewable Energy Agency (IRENA) also estimates that by 2050 up to 50% of the final energy consumption will be in the form of direct electrification, with renewables contributing up to about 80% of the final energy consumed [11]. This would entail installing very large numbers of wind farms each year between 2020 and 2050, involving huge investments. Given the relatively low life cycle GHG emissions of nuclear power, accelerating its adoption would also enable stretching the carbon budget further by displacing emissions from sectors that are or that can be readily electrified, leaving some room to accommodate emissions from the hard to abate industries (e.g. cement production, chemicals synthesis) without overshooting the budget until viable alternatives are found for their decarbonization. The IAEA estimates that nuclear power avoided a total of 74 Gt CO_2 over the period 1971–2018, equivalent to the cumulative emissions from the entire power sector for the six years from 2013 to 2018 [12].

Variable renewable energy systems such as solar PV systems and wind turbines primarily produce clean electricity with regional, diurnal and seasonal fluctuations (with average plant load factors of ~30%, when energy storage is not considered). Nuclear power plants, on the other hand, can reliably supply heat as well as electricity with very low life cycle GHG emissions [12], at stable costs and low marginal prices and with very high plant availability factors (~90% per year for most water cooled reactors operational today) while demonstrating high levels of resilience under a wide range of ambient conditions and weather events. They are also highly concentrated energy systems compared with dilute, variable renewable energy technologies, and thus the land area requirement is significantly lower for nuclear power projects compared with other low carbon energy systems (estimated to be about 100–150 times less in the European Union compared with solar PV for identical units of electricity generated on an annual basis [13]). Additionally, the number and quantity of critical raw materials and mining-intensive minerals are much lower for the nuclear power sector than for the renewable energy sector, which makes it much more resource efficient and much less vulnerable to several supply chain disruptions and other geopolitical factors [14]. Wider deployment of nuclear reactors would be needed to complement renewables and ensure an affordable and accessible energy transition for all.

2.2. APPLICATIONS OF HYDROGEN IN A DECARBONIZED WORLD

Energy use is not the same as electricity use. Globally, about 20% of the final energy consumption today is in the form of electricity, and this is slated to rise further as more end use sectors undergo electrification to attain net zero emissions [15]. To date, however, most of the activity in the energy transition space has been aimed at increasing the deployment of renewables, which would primarily contribute to the decarbonization of the electricity sector. While this is an important step towards a sustainable energy transition, given that the electrification of many sectors is at the heart of every net zero strategy, there are sectors that will not be decarbonized by clean electrification alone. Where direct electrification is not a feasible decarbonization pathway (e.g. in heavy transport and long haul transport, aviation or shipping; in industrial heat applications) or where a carbon free feedstock or reducing agent is required in a process technology (e.g. iron ore reduction, steel or ammonia production), hydrogen produced from a carbon free starting material such as water using low carbon electricity is a viable non-fossil alternative that is gaining increasing acceptability in industry [16, 17]. Using either renewable or nuclear electricity with mature technologies such as water electrolysers, low carbon hydrogen can be produced in bulk quantities (further details and other potential synthesis routes are discussed in Section 4).

The many possible uses of low carbon hydrogen may be classified into essential, conditional and avoidable, as shown in Fig. 1. The categorization is not meant to be exhaustive, but rather an exemplification based on the conceptual framework given in Refs [18, 19]. This categorization may change over time, depending on the supporting policies in place and expanding markets.

The important application areas for low carbon hydrogen are the following:

— Replacement of existing conventional feedstock. Hydrogen can replace fossil fuels as energy sources and feedstocks (particularly coal and natural gas) in multiple sectors [20, 21]. Hydrogen may be used directly as the energy vector in many of these applications, or it may be converted into synthetic fuels such as methanol (using post-combustion CO_2 captured from thermal power plants or cement plants, or even directly from air) or

Essential	Conditional	Avoidable
Ammonia/fertilizers Food processing Iron, steel, other metallurgical uses Hydrotreatment/ hydroprocessing applications in refineries Methanol synthesis Shipping/aviation/ long distance/heavy transport	Long distance rail transport River/lake based local water transportation Industrial heating/ thermal applications Autonomous power systems in remote areas/islands	Renewable energy long duration storage/ transportation Buses, cars and other modes of urban transport Domestic heating, cooking and other energy services Grid balancing/ flexibility/reliability services

FIG. 1. Classification of the potential applications of low carbon hydrogen.

ammonia (using nitrogen separated from air) or into a variety of liquid and solid hydrocarbons by the Fischer–Tropsch synthesis process, based on catalytic reactions of carbon monoxide or CO_2 and hydrogen [22]. The liquid fuels can serve as drop-in substitutes for fossil fuels, requiring minimal infrastructural changes on the end user's side.

— Integrated energy systems with multipurpose applications. The production and storage of hydrogen using excess or surplus low carbon electricity (particularly generated by wind and solar PV) has been suggested as an effective means for longer duration (~1 week or more) energy storage [23], as well as for energy arbitrage to take advantage of differential power tariffs at different times of the day [24]. This may be an alternative to grid scale battery electricity storage, particularly where geographical or geological features such as underground caverns or salt caverns are available for large volume gas storage. In such an arrangement, the stored hydrogen would be converted back into electricity through grid scale fuel cell stacks and re-exported to the grid at times of higher power demand. Owing to its high calorific value, hydrogen has also been envisaged as a source of low carbon industrial heat using combustion either on its own or initially blended with natural gas in industrial furnaces [25].
— Steel production. Hydrogen as the heating medium and reducing agent for iron ore in steel production has the potential to drastically reduce the carbon footprint of the industry by reducing and ultimately eliminating the use of coke or natural gas [26]. Low carbon steel production has been extensively investigated by researchers (e.g. Ref. [27]), and practical demonstrations of the technology have also been accomplished, and are set to scale up in the next few years.
— Ammonia production. Producing ammonia using low carbon energy sources and hydrogen has attracted attention worldwide, and industrial consortiums are being set up to establish the first ventures to manufacture, transport and use low carbon ammonia. Ammonia is the raw material for the fertilizer industry and may also be considered as a shipping fuel and hydrogen carrier.
— Power production. Gas turbines working on natural gas combustion are being retrofitted to work with blends of natural gas and low carbon hydrogen, with the ultimate objective of developing 100% hydrogen turbines for power production [28]. These turbines may replace natural gas turbines in existing gas fired power plants and become practical examples of power-to-gas-to-power systems, as a complement to fuel cell based power packs.
— Mobility. For long haul and heavy transport, including aviation and shipping, hydrogen and hydrogen derived products (such as synthetic fuels) are being considered as sustainable alternatives to fossil fuels, since battery based electrification may not be entirely feasible for these applications [29]. The

high battery weight and long battery charging time for these applications using present-day technologies suggest that an energy dense, low carbon or nearly zero carbon fuel with quick refuelling time is crucial. Hydrogen (and its derivatives) is one such alternative. Three areas where its use is currently being considered are provided:

(1) Surface mobility. In the surface transport sector, hydrogen fuel cells could be used in trucks and buses, with the accompanying need to set up hydrogen refuelling stations along major motorways. In some areas, hydrogen based trains may also be a way to replace fossil fuels such as diesel in the railway sector.

(2) Water transport. Hydrogen could be used as an alternative to marine diesel and heavy fuel oils in boats, ferries and ships for decarbonizing water transport. For shorter distances covered by boats, ferries and recreational vessels on inland waterways, direct electrification is also feasible, but for shipping over long distances and freight transport by ships, the use of hydrogen or its derivatives appears to be much more viable than electrification. Liquid hydrogen and ammonia are practical options. Development of high power fuel cell stacks for marine transport applications is being carried out by some organizations [30, 31]. Hydrogen bunkering arrangements (particularly for liquid hydrogen) will have to be created as part of a hydrogen friendly port infrastructure.

(3) Air transport. Hydrogen powered aircraft as well as the production and upgrade of sustainable aviation fuels from biomass are options for decarbonizing aviation. Both cases would need enormous quantities of hydrogen, along with a hydrogen bunkering infrastructure at airports in the former case.

Distributed production of hydrogen would require the establishment of a hydrogen transmission infrastructure connecting producers and consumers. One of the ideas being explored in many countries is to repurpose existing natural gas pipelines and compression stations for hydrogen transfer. Major technology upgrades are not expected to be necessary for a blend of 5–20% hydrogen with natural gas, making this a possible near term use of hydrogen. However, the long term advantages with respect to decarbonization are less certain [32, 33]. When the hydrogen produced is intended for ammonia production, it may be more feasible to transport it by ship as liquid ammonia, especially for bulk quantities over continental scale distances.

2.3. ESTIMATES OF HYDROGEN DEMAND FOR NET ZERO

2.3.1. Current hydrogen requirements and projections for the future

Current global hydrogen consumption is estimated to be over 70 million tonnes (Mt) per year (in refining and ammonia production), with the demand in 2070 projected to be over 500 Mt per year in the power, transport, industrial and chemical sectors [34].

The IEA estimates that to decarbonize hydrogen-consuming sectors and support upcoming hydrogen consumers, globally about 850 GW(e) of installed water electrolyser capacity would be needed by 2030 and 3600 GW(e) by 2050 to produce sufficient low carbon hydrogen for deep decarbonization of fossil fuel dependent sectors [10]. Other estimates, such as those by IRENA, also suggest that by 2050 annual global hydrogen demand would be about 74 EJ (corresponding to 530 Mt of hydrogen per year when produced by water electrolysis) [35]. According to Ref. [36], 3–12% of final electricity consumption in a net zero world may have to be diverted to electrolytic hydrogen production, depending on the availability and efficiency of alternative technologies such as carbon capture, particularly with respect to hard to abate sectors.

2.3.2. Role of nuclear enabled hydrogen production

While renewables have been identified as the source of electricity for low carbon hydrogen production in many national hydrogen strategies, in nuclear equipped countries a part of this clean hydrogen demand (about 3–5%) can also be met by coupling water electrolysers with existing or upcoming nuclear power plants operating in cogeneration mode [10]. Providing a stable baseload performance, nuclear power plants are well suited for reliably supplying the large quantities of clean hydrogen needed for attaining net zero emissions in various critical industries. With respect to the hard to abate sectors that would use the hydrogen produced, nuclear hydrogen production processes represent an important application of nuclear power with a significant contribution to climate commitments, beyond serving the electricity sector and its decarbonization agenda.

Despite the well recognized role of hydrogen in decarbonization targets aligned with the Paris Agreement goals [37, 38] and the significant momentum surrounding the creation of a hydrogen economy, the realization of these ambitions through deployment of projects at scale is still slow. At current growth rates of low carbon hydrogen facilities, globally only about 0.5–5% of final energy consumption may be for hydrogen production by 2050, which is less than a third of the actual requirements in most net zero pathways [39]. Thus to bridge this gap, accelerated deployment of hydrogen production and its uptake by end use sectors has to be made a priority area for policy makers.

Globally, 40 national hydrogen strategies had already been put forward as of September 2022 [2], and the number continues to grow, driven by net zero goals. Some of these strategies include projects on nuclear enabled hydrogen production, but most low carbon hydrogen projects are based on renewables. Additional details about country specific programmes are available in the Annex to this publication.

2.4. CURRENT ROUTES TO HYDROGEN PRODUCTION

Currently, the most widely used commercial technology to produce hydrogen is steam methane reforming. In some cases, naphtha reforming is also used in a similar manner. Most petroleum refinery complexes operate their own on-site reformer reactors, producing thousands of kilograms of hydrogen on their premises; this could be converted into lower carbon intensity hydrogen if carbon capture technologies were also deployed. The techno-commercial features of carbon capture technologies are still not mature, and as yet there is no clarity on their deployment rates. Typically, about 8–9 t of CO_2 are released per tonne of hydrogen produced by this technology [40]. However, this does not take into account the additional emissions of GHGs in the natural gas value chain (i.e. fugitive methane releases during natural gas extraction, processing and distribution, which have a global warming potential higher than that of an equal mass of CO_2) [41].

At present, the cost of hydrogen produced from the steam methane reforming process is estimated to be between US $1.50 and US $3.50 per kg of H_2, depending on the prevailing natural gas cost (which has shown significant volatility and supply side concerns in recent times), the hydrogen production scale or plant size and whether there is consideration of any carbon capture process or any form of carbon pricing prevailing in the region or area [42]. Currently available carbon capture technologies (with carbon removal efficiencies of about 85–90% and costing between US $150 and US $300 per tonne of CO_2 captured) would raise hydrogen costs by US $1.20–2.70 per kg of H_2 from steam methane reforming plants [43]. Thus, under these conditions, hydrogen from steam methane reforming plants with carbon capture, utilization and storage (CCUS) or carbon capture and storage (CCS) would cost about US $3.50–6.20 per kg of H_2, not considering any carbon taxes paid on the residual 10–15% of unmitigated emissions from the carbon capture plant itself.

An attempt to estimate the life cycle CO_2[1] and methane emissions in the production of hydrogen by steam methane reforming, with and without CCUS/CCS, is provided in Ref. [41]. The study uses a fugitive methane emission rate of 3.5% of total natural gas handled and the 20 year global warming potential in one of the scenarios. It concludes that total GHG emissions from hydrogen production with steam methane reforming and on-site carbon capture will only be 9–12% lower than that from an equivalent hydrogen plant based on steam methane reforming without CCUS/CCS, as additional natural gas consumption is needed to supply the energy requirements for running the carbon capture facilities at steam methane reforming based hydrogen plants with CCUS/CCS.

It has to be noted that any existing and newly created hydrogen infrastructure based on hydrocarbons faces premature closure as well as stranded asset risk and carbon risk due to possible changes in environmental and financial regulations (e.g. rapidly rising carbon taxes or prices in emissions trading schemes), not to mention energy security considerations in countries that are heavily dependent on gas imports. The other aspect is that carbon capture projects are extremely capital intensive, and the economic, environmental and social cost–benefit considerations are still not clearly demonstrated, making these projects still very risky for developers and financers. Low carbon prices in most jurisdictions have been identified as one of the factors responsible for project failure, and this has prevented their effective deployment at scale [46].

Other potential routes to large scale hydrogen production are biomass gasification [47] and coal gasification [48]. Both of these techniques can yield low carbon hydrogen when coupled with carbon capture technologies, but there would still be residual emissions due to the inherent inefficiencies of capture. The permanence of the capture also has to be ensured and verified independently. Methane pyrolysis using plasma techniques is another viable option for countries rich in natural gas resources, as the only products of this process are hydrogen gas and solid carbon [49].

For the near term, catalytic pathways (using high temperature fixed bed chemical reactors with solid catalysts or bubble columns with liquid catalysts) are the simpler option and may be at a higher TRL than the plasma based route, but they are still less developed than the electrolytic routes or hydrocarbon reforming routes [50].

---

[1] Life cycle CO_2 emissions refers to the CO_2 emissions covering all the activities (particularly energy use) associated with a product or technological solutions, from extraction of raw materials to its manufacture and actual operation over its design life to its dismantling and decommissioning at the end of its life. The internationally accepted standards for the methodology to estimate life cycle impacts, including carbon emissions associated with a product or a process, are ISO 14040 [44] and ISO 14044 [45].

2.5. OPPORTUNITIES, CHALLENGES AND UNCERTAINTIES OF THE HYDROGEN ECONOMY

The hydrogen economy would be a crucial part of a global low carbon transition, but despite the numerous opportunities and the corresponding technological advances already made in several individual facets, there are still system level challenges that need to be overcome. Some further insights are provided in this section, based on Refs [51, 52].

2.5.1. Opportunities

The following opportunities for the hydrogen economy have been identified:

- Production. The production and use of low carbon hydrogen opens new possibilities for sector coupling and hence integrated energy systems planning, fostering systems level thinking instead of the prevalent siloed approach to dealing with various primary energy sources and final energy use patterns.
- Drop-in replacement potential. The use of low carbon hydrogen is a crucial step in the decarbonization of several existing industries, such as petroleum/petrochemicals production and ammonia synthesis, and can reduce their dependence on natural gas (particularly imported natural gas). Hydrogen produced from low carbon energy sources is a drop-in substitute for fossil based hydrogen in these sectors. They represent the early users of any form of non-fossil-fuel derived hydrogen, as no major changes in their hydrogen end use system will be required for switching from one source of hydrogen to another.
- Manufacture of equipment. For successful implementation of the hydrogen economy as a part of net zero strategies, end use technologies for hydrogen — such as fuel cells, hydrogen gas turbines and furnaces for direct reduction of iron (DRI) using hydrogen — have to attain technical and commercial maturity. Whereas some of the technologies such as DRI are at the pilot scale demonstration level, others such as fuel cells have attained a much higher TRL. This would be crucial for ensuring ready offtake of the produced hydrogen and a stable business model over the long term, which would be important for ensuring the techno-commercial viability of the projects. Accelerated manufacture of the hardware components of a hydrogen economy needs to be prioritized in keeping with the long pipelines of projects already announced all over the world.
- Water electrolysis. Water electrolysers and fuel cells of various designs and capacities are available globally, with several vendors for each of these

technologies. This creates a competitive market environment, facilitating further cost reductions and benefits from the technology learning curve effects. The suppliers of other components such as high pressure hydrogen storage systems, gas compressors, sensors, instruments and purification systems are also growing in number, as is the diversity of products offered. However, there is still scope for technical improvements and cost reductions for most of these components, which presents opportunities for the manufacturing sector.

2.5.2. Challenges

The following challenges for the hydrogen economy have been identified:

— Electricity need. The first challenge is to develop the dedicated clean electricity infrastructure to produce vast quantities of hydrogen and then to distribute the hydrogen to users, since not all the hydrogen can be produced on-site or close to user locations. Nor can there be large scale hydrogen storage facilities at all potential user locations owing to space constraints and potentially to safety and regulatory considerations.
— Water availability. Currently, many viable sites for renewable electricity production are in remote, arid or semi-arid regions. While renewable electricity may become very inexpensive at these locations, the sites might not necessarily be suitable for hydrogen production owing to constraints on water availability. In this regard, nuclear power plants have a distinct advantage for collocation with hydrogen production facilities, since ensuring long term water availability is one of the most important siting criteria in the deployment of nuclear power programmes.
— Critical mineral supply. Securing reliable supply chains for critical minerals and materials relevant to the hydrogen economy (e.g. nickel, platinum group metals for use in water electrolysers, electro-catalyst materials) may become a significant challenge as the hydrogen economy ramps up.

2.5.3. Uncertainties

The hydrogen economy is a crucial component of a net zero world, with many beneficial impacts on the environment. However, there are potential negative environmental aspects that have to be taken into consideration and examined further. Some issues uncovered in a few recent studies [18, 53] are the following:

- Leakage and fugitive emissions of hydrogen (which could be anywhere between 0.02% and 13.2% of the total hydrogen production, distribution and use, according to some estimates, as included in Ref. [18]) have the potential to create radiative forcing and global warming, as does the water vapour that will be produced in most end uses that involve its combustion (direct or electrochemical).
- A chemistry–climate model-based study estimates that in a global hydrogen economy, the hydrogen surface mixing ratio could increase by 50–300% from the current value of 0.5 ppm (depending on uncertainty in release rates), with corresponding changes in atmospheric concentrations of hydroxyl radicals, ozone levels, nitrogen oxides and water vapour [53]. While water vapour has a short atmospheric lifetime and is not expected to persist long enough in the lower atmosphere to create a substantial radiative forcing or warming effect, fugitive emissions of hydrogen (due to permeation through materials, leakage, venting, purging and/or boil-off losses) from its value chain can have a significant warming impact, with estimated forcing of up to 0.148 $W m^{-2}$ for an atmospheric hydrogen level rise of 1.5 ppm. Another study estimates the global warming potential of hydrogen to be 11 times greater than that of CO_2 over a 100 year horizon [18]. Thus, hydrogen emissions accounting and monitoring are important, which implies that hydrogen sensor and detection technologies need to make significant advances if they are to be used widely.

2.6. MACROECONOMIC IMPACTS OF COMMERCIAL HYDROGEN ENERGY SYSTEM DEPLOYMENT

A large scale energy transition, including widespread electrification and use of low carbon hydrogen, is naturally expected to create economy wide impacts at regional, national and global levels. Some of these potential impacts are as follows [54, 55]:

- Energy security. Several countries may be able to use domestic energy resources, including nuclear assets, to support clean hydrogen production, therefore improving national energy security and reducing import bills for energy commodities such as natural gas and crude oil or liquid hydrocarbons for use in various sectors.
- Clean fuel exports. Countries with a large generation capacity and low domestic demand for hydrogen can become net exporters of energy in the form of hydrogen and its derivatives (e.g. ammonia, methanol, other synthetic fuels) to countries where land, water or resource constraints will

not enable enough hydrogen to be produced locally. This is expected to create new energy trade avenues and shifts in the geopolitics of energy. This will require internationally harmonized codes, standards and certification schemes for low carbon hydrogen.

— Carbon credits. It has been suggested in some assessments that deployment of a hydrogen economy will reduce the need for carbon emissions trading and carbon offsets in some sectors, thus freeing up residual offsets for the hard to abate industries that do not yet have viable decarbonization routes [56]. The earnings from these offsets could be partly or fully reinvested in advancing clean hydrogen projects.
— Carbon benefits. Several studies have established that investments in low carbon energy systems (i.e. renewables, nuclear) create multiple economic multipliers that amplify the benefits from the deployment of these systems [57]. The hydrogen economy, which is a derivative of low carbon energy systems, may be expected to further improve this multiplier effect.
— Asset protection and revitalization. Accelerated deployment of clean hydrogen projects can contribute to the early retirement or phase-out of fossil infrastructure, leading to stranded assets worth large amounts of money and to the loss of fossil fuel related taxes and surcharges (e.g. sales tax, which is a major source of government revenue in many countries).
— Generation of new employment opportunities. The emerging hydrogen and fuel cell industries would require a wide variety of occupations at all skill levels, inducing new opportunities for direct and indirect jobs for technology deployment, supply chain and other related activities.

3. COMMERCIAL DEPLOYMENT ASPECTS

Development and deployment of commercial nuclear hydrogen programmes depend on careful consideration of several factors that are relevant to national governments, nuclear power plant owner/operators, technology suppliers and hydrogen end users. This section presents a comprehensive picture of all the commercial deployment aspects of nuclear hydrogen production that will contribute to the creation of a successful and replicable business case for these projects.

3.1. OPPORTUNITIES FOR THE NUCLEAR INDUSTRY IN THE HYDROGEN MARKET

3.1.1. General overview of the hydrogen market

Current global hydrogen consumption is about 95 Mt per year [58]. Over the past fifty years, hydrogen consumption has grown steadily by 1.6% per year (see Fig. 2). Various reports provide a range of scenarios for 2050 [35, 58–60], all of which foresee much faster growth in demand over the coming decades.

The largest share of the hydrogen produced today, mainly from fossil fuels, is used in refineries or for chemical production [52, 61]. As shown in Fig. 2, the global push to move towards a net zero target would increase the hydrogen demand, for example, due to its application as fuel. To play an essential role in deep decarbonization scenarios, hydrogen has to be produced by low carbon energy sources such as nuclear. The IEA estimates that global hydrogen demand will reach 530 Mt per year by 2050, in a net zero scenario [51].

3.1.1.1. Current market for hydrogen

Today, most hydrogen is produced near the consumption site owing to the high transport and storage costs [62]. Currently, industrial applications are the main markets for hydrogen. The largest industrial application of hydrogen is in refineries, accounting for 40 Mt (42% of the total) [58], where it is mainly used for hydrodesulfurization and production of light hydrocarbons from heavy

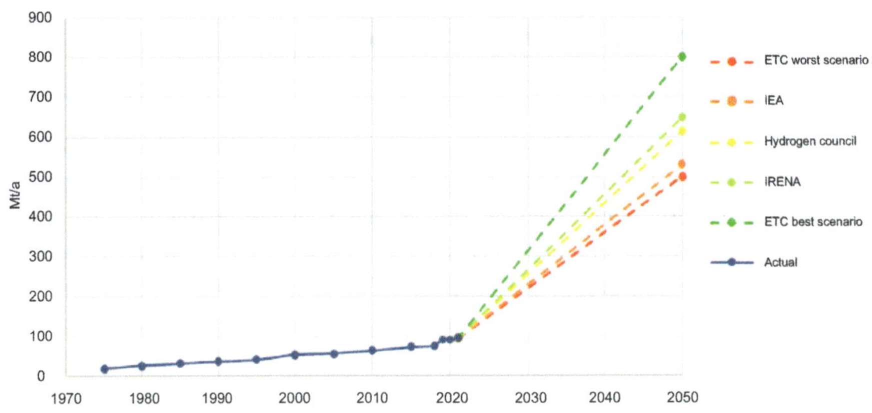

FIG. 2. Global hydrogen consumption between 1975 and 2021, with predicted demand to 2050 (based on data from Refs [35, 58–60]).

hydrocarbons. The second largest sector is ammonia production (34 Mt, 36% of the total) [58], where hydrogen is used for the fixation of nitrogen from the air. Ammonia is the main feedstock for several sectors, such as nitrogen based fertilizers, refrigerant gases and pharmaceuticals. Methanol production is the third largest sector, consuming 15 Mt of hydrogen per year (16% of the total) [58]. Methanol is currently used to produce plastics and fuel additives. Finally, DRI accounts for 5% of the total hydrogen consumption (5 Mt per year) [58]. DRI is a process where iron ore is reduced to produce sponge iron, which will later be transformed into steel. In this sector, hydrogen is used as a reducing agent. Other sectors that use hydrogen are transport and residential heating, but these account for less than 1% of global consumption [58]. Figure 3 presents the numbers for the hydrogen market as of 2022 (based on data from Ref. [58]).

The biggest hydrogen producers are shown in Fig. 4, based on data from Ref. [54]. These countries and geographical regions are also the biggest hydrogen consumers, as less than 0.3% of the total hydrogen consumed is traded in the global market [63] because it is less expensive to transport and store natural gas (the main hydrogen feedstock). The distribution of hydrogen consumption is non-homogeneous across countries. The four main producers/consumers of hydrogen alone account for approximately 56% of the hydrogen used yearly. China alone accounts for 27% of the consumption, the United States of America for 13%, India for 8% and the Russian Federation for 7% [54]. In addition, these four countries have active and well-developed nuclear programmes that can support decarbonization of their industries, including hydrogen production.

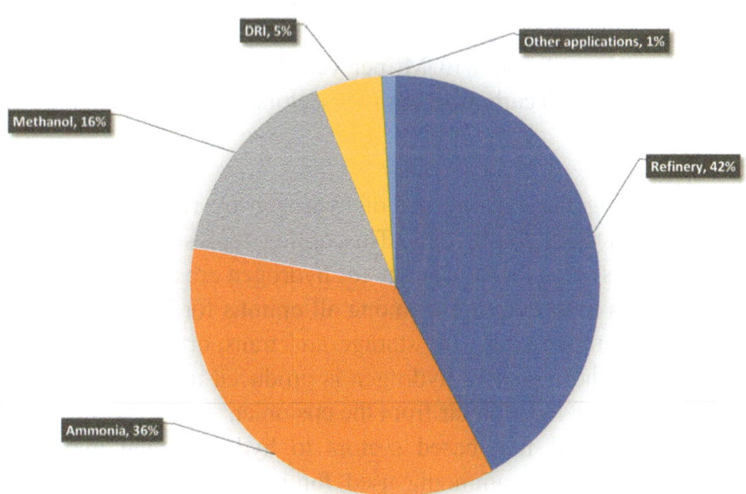

FIG. 3. Hydrogen market, as of 2022 (based on data from Ref. [58]).

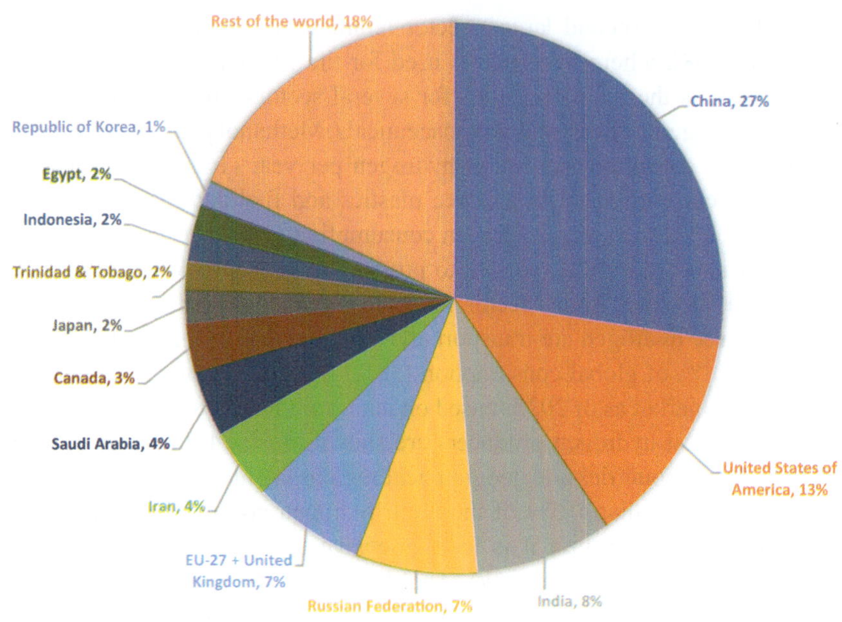

FIG. 4. Hydrogen consumption by 2050 by country/region (based on data from Ref. [54]).

3.1.1.2. Future markets for hydrogen

Several scenarios elaborated by the Energy Transitions Commission (ETC), Hydrogen Council, IEA and IRENA [54, 58–60] show future hydrogen demand in the range of 500–813 Mt per year by 2050, reflecting the amounts required to massively decarbonize various sectors. Its applications are expected to be the following:

- Decarbonizing the existing use of hydrogen by replacing the production of hydrogen from fossil fuels. This option represents short term market opportunities since the consumers for hydrogen are already established and the option is most accessible among all options for hydrogen applications. Additional infrastructure for storage and transportation might be needed in the case where nuclear hydrogen is produced using an existing nuclear power plant at a location far from the customer.
- Converting fossil fuel based sectors to hydrogen and hydrogen based synthetic fuels (e.g. ammonia used for internal combustion engines for large ships). Low carbon hydrogen can be used directly in fuel cells or as a feedstock to produce synthetic fuels that could support sectors such as

long-haul road and maritime transportation. In this case, new infrastructure would be needed to convert hydrogen into synthetic fuels and efficient fuel cells would have to be manufactured. This option represents midterm market opportunities, since new upstream infrastructure and fuel cell technologies are needed.

— Decarbonizing other sectors through low carbon hydrogen penetration. New infrastructure to use hydrogen would be needed in this case, for example to use hydrogen in residential heating. This option represents long term market opportunities, since new upstream (production), midstream (distribution) and downstream (consumption) infrastructure is needed.

More generally, Fig. 5, derived from data from the ETC [59] and the IEA [51], shows the most promising sectors foreseen for 2050. Fertilizers (using hydrogen for producing ammonia) will use 38 Mt of hydrogen per year, methanol will use 36 Mt per year, refining will use 11 Mt per year and steel-making will use 87.5 Mt per year [51, 59]. Similar to the current market, fertilizers (ammonia), methanol, refining and steel production will continue to be the four markets with relevant hydrogen demand. The opportunity for nuclear hydrogen in these markets is to replace the existing hydrogen produced from fossil fuels, which

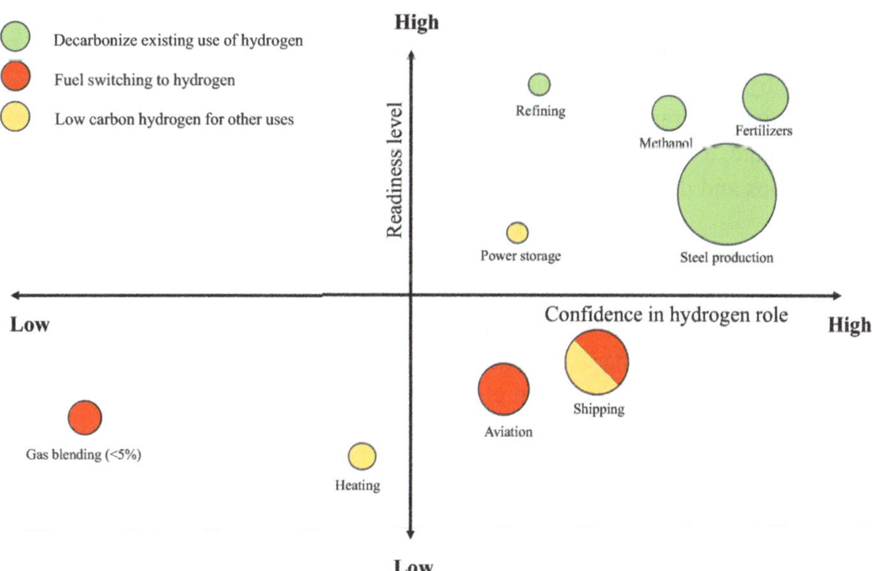

FIG. 5. Potential uses of hydrogen in a low carbon economy by 2050 (based on data from Refs [51, 59]).

would require substantial infrastructural changes. However, since hydrogen (from fossil fuel) is already used in these markets, the transition to low carbon hydrogen is expected to be easier than developing a completely new supply and demand infrastructure. The change is likely to occur upstream in the value chain, where hydrogen will be produced with low carbon sources replacing fossil fuels. Accordingly, the barriers to low carbon hydrogen in these markets are low, allowing reasonably quick entry.

According to the ETC [59] and the IEA [51], power storage, shipping and aviation represent future markets that will need heavy reconfiguration to foster hydrogen use. The future hydrogen demand for these sectors is 84 Mt per year for power storage; 141.5 Mt per year for sea shipping (which refers to the overall amount of pure hydrogen required; however, consumption could also include hydrogen based synthetic fuels); and 97.5 Mt per year for aviation (using hydrogen based synthetic fuels) [51, 59]. The technologies necessary for these markets are different and clear decarbonization pathways are yet to be defined. For example, shipping industries could be powered by several hydrogen based fuels using internal combustion engines, pure hydrogen using fuel cells and/or electric hybrid solutions. These new markets offer clear opportunities for hydrogen to penetrate traditional fossil fuel based markets. Therefore, whereas the current market uses hydrogen only as a chemical compound to obtain a specific chemical reaction, future markets would see hydrogen as both a chemical compound and as an energy carrier to couple low carbon energy (e.g. nuclear) with other sectors. Finally, gas blending and residential heating are each expected to require 15 Mt per year by 2050. However, gas blending might be useful only in the short term decarbonization transition, while residential heating using pure hydrogen is still under debate, as it would compete against effective existing alternatives such as heat pumps and district heating [64].

3.1.2. Nuclear potential and value highlighted in the hydrogen markets

Compared with other low carbon intermittent energy sources (e.g. wind, solar), nuclear power has the advantage of constantly producing a large amount of energy, generating both clean electricity and heat. At the same time, the low fuel cost, the distribution of capital costs over continuous operation, and the generation of an additional commodity (in this case, hydrogen) beyond electricity can be financially advantageous for the nuclear plant itself and can reduce the cost of hydrogen production.

Some potential hydrogen markets would possibly require great amounts of hydrogen concentrated in relatively small areas. Consequently, some areas would start to develop large hydrogen consumption centres. For example, bunker ports that buy, store and sell bunker fuel could be collocated with nuclear facilities that produce, store and sell hydrogen. Owing to the advantages of nuclear power, nuclear hydrogen presents opportunities within these large consumption centres.

Small modular reactors (SMRs) could be deployed to decarbonize aviation, shipping, cement manufacturing and other industries by mid-century, in a cost competitive manner. These hard to abate economic sectors could be decarbonized with the large scale use of hydrogen as a clean energy carrier and as a feedstock for synthetic fuels such as ammonia, synthetic diesel and jet fuel. To achieve this, hydrogen enabled fuels need to be produced, without emissions, at prices that are competitive with those of the fossil fuels they are replacing. Hydrogen production can be considered from the very early stages of SMR design — this would improve the economic competitiveness of SMRs, given the higher synergies that can be realized with their flexibility and modularity.

Figure 4 indicates relevant differences among countries and regions regarding hydrogen consumption; the experience in building and operating nuclear infrastructure varies across countries as well. There are countries with operating nuclear power plants, countries with plants under construction, countries with planned nuclear plants, and countries with no interest in nuclear power technologies. Therefore, to assess the role that nuclear energy would play globally to support hydrogen production, it would be useful to identify countries that are in favour of the development of nuclear hydrogen.

Figure 6 considers two variables, nuclear commitment and potential market size. On the ordinate axis, the nuclear commitment is a combination of many parameters (e.g. the number of reactors active, planned or under construction; the percentage of the energy mix of nuclear or gigawatts installed; the number of nuclear energy producing countries and the non-nuclear energy producing countries with strong pledges to introduce nuclear capacity). The positioning on this axis can indicate a potential time horizon in which nuclear hydrogen can start playing a significant role. On the abscissa axis, it is possible to place the current

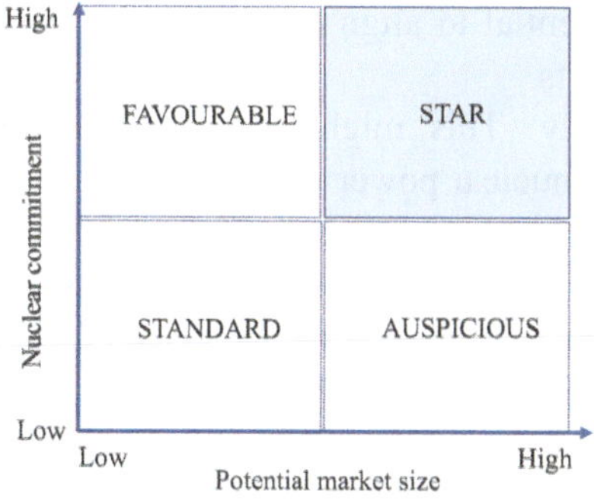

FIG. 6. Nuclear hydrogen market identification framework.

or potential hydrogen volumes used in a national market. The positioning on this axis can represent an indicator of the attractiveness of a market for nuclear hydrogen. Therefore, there are four macro categories within this Cartesian space: standard, favourable, auspicious and star. Countries with high nuclear commitment and a big hydrogen market would be more likely to produce nuclear hydrogen in the short term. Conversely, countries with low nuclear commitment and a small hydrogen market would be more likely to use other sources for hydrogen production, at least in the short and medium term.

Value in general includes both tangible (e.g. increasing revenues, savings in costs and time) and intangible benefits (e.g. improving quality, developing corporate competencies, cultivating personnel, improving the satisfaction of the stakeholders, protecting the environment) [65]. For example, building a plant capable of producing hydrogen from nuclear energy creates new jobs and increases the contractor's revenues. At the same time, the contractor acquires knowledge that may become a source of competitive advantage when developing new infrastructures. It is useful to assess the value creation and capture mechanisms within a project life cycle, considering all the actors involved and their relationships [66]. Industrial players may take advantage of developing projects that create value over time for multiple stakeholders. In summary, the value that nuclear hydrogen can generate goes beyond economic and financial performance, and includes the following:

- Low carbon emission: Nuclear is a competitive energy source to produce low carbon hydrogen at a large scale. The IAEA's net zero scenario highlights the need for the coexistence of variable sources (renewables) and stable sources (nuclear) [67]. This represents a competitive advantage for nuclear energy, compared with fossil fuels, even when the latter is coupled with carbon capture and storage.
- Production schedule: Nuclear hydrogen production can be scheduled, whereas renewable production cannot. The possibility of planning hydrogen production is essential to align demand and supply to reduce storage and transportation costs.
- Density of supply: This might facilitate the collocation of hydrogen production using nuclear power with large consumption centres.
- Price stability: Nuclear hydrogen has less exposure to price volatility than fossil fuels. Furthermore, nuclear power plants are capital intensive, and almost all their costs are fixed or sunk costs. From this perspective, nuclear hydrogen guarantees price stability over time.
- An alternative energy product: The ability to switch from electricity to hydrogen production (e.g. during periods of low demand, when hydrogen can be stocked) is a valuable option for both renewables and nuclear energy.

In addition, using heat from nuclear power plants to produce hydrogen brings the advantages of increasing efficiencies in hydrogen production, increasing energy efficiency of the nuclear power plant and reducing the nuclear heat load on the environment.

— First mover advantage: Nuclear hydrogen can represent an opportunity for countries to be the first mover on the market. Developing new competencies and know-how can allow countries to export their technology worldwide.

Nuclear hydrogen can also generate value outside the nuclear industry. The availability of large quantities of hydrogen at low, stable prices can incentivize new actors to invest in this technology. For example, the heavy transport industry (e.g. trucks, shipping) could invest in hydrogen as a substitute for fossil fuels. Similarly, universities could use this opportunity to broaden their research fields and train new professionals. The involvement of new secondary stakeholders could positively affect the market, generating new employment opportunities. In this case, secondary stakeholders are individuals, groups of people or organizations that can be affected or are perceived to be affected by the hydrogen value chain but that are not strictly linked through contractual relationships with the value chain itself. Secondary stakeholders include universities, consumers and local communities [68]. Therefore, the involvement of these actors can also improve companies' reputations by leveraging new energy sources for achieving lower CO_2 emissions.

However, because the value generated by nuclear hydrogen can be judged by different stakeholders according to their needs, stakeholders have to be identified and mapped in the value chain. To engage stakeholders, it is therefore necessary to understand the value that nuclear hydrogen can generate for all of them. Here are a few examples of potential value chains for nuclear hydrogen:

— Ammonia value chain: Ammonia is used in the production of nitrogen based fertilizers and refrigerant gases, and in the pharmaceutical industry. As an example, a simplified potential value chain of nitrogen based fertilizers is shown in Fig. 7, starting from the production of nuclear hydrogen. In this value chain, the main steps are the production of hydrogen from nuclear energy, the chemical transformation of hydrogen into ammonia, the production of fertilizers and the consumption of these fertilizers by end users. Table 1 presents values created for different stakeholders involved in the value chain of nuclear hydrogen for the fertilizer market.

TABLE 1. VALUE CREATION FOR DIFFERENT STAKEHOLDERS INVOLVED IN THE CHAIN OF NUCLEAR HYDROGEN FOR THE FERTILIZER MARKET

Component	Stakeholder	Value
Nuclear power plant	Nuclear power plant owner	New investments, market growth, diversification, option for cogeneration
Ammonia plant	Ammonia producer	Hydrogen price stability, security of supply, contribution to climate commitments
	Ammonia plant owner	Growing market, new locations
Fertilizer factory	Nitrogen based fertilizer producers	Ammonia price stability, security of supply, contribution to climate commitments
	Environmental groups and non-governmental organizations	Lower environmental impact, CO_2 emission reduction
End users	Farmers	Decreased uncertainty, security of supply, price stability, contribution to climate commitments
	Agricultural associations	
	Agricultural consortia	

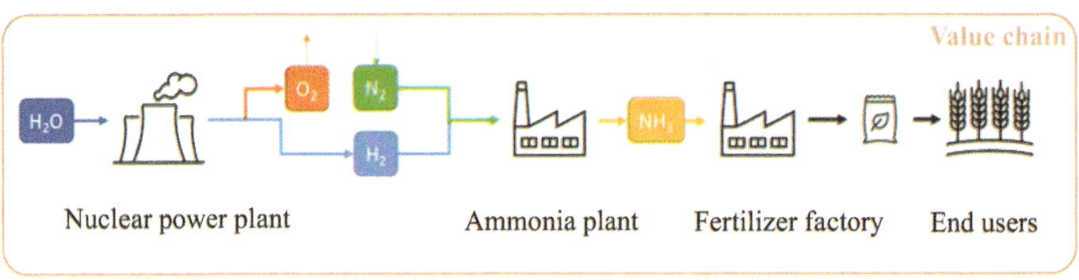

FIG. 7. Nuclear hydrogen fertilizer value chain.

— Maritime transport value chain: The value chain for using nuclear hydrogen as a fuel for maritime transport (see Fig. 8) is a potential future market for hydrogen. Hydrogen could be produced within bunker ports and then used as fuel for heavy shipping vehicles to limit the cost of transport. Placing the nuclear power plant within the bunker port is ideal because of the availability of cooling water and water for electrolysis. The power plant would also supply the needed energy to the bunker port. The scenario is particularly attractive when fresh water is available close to shipping areas. If only sea water is available, desalination units would be necessary. It would be necessary to liquefy and store hydrogen in specific vessels also located within the bunker port. Then, the liquified hydrogen could be transported through bunkering ships to the end user. In this supply chain, the value creation for the different stakeholders involved is illustrated in Table 2.

— Direct reduction of iron: Steel is one of the materials with the largest production volume. Figure 9 represents a simplified steel-making value chain where nuclear power is adopted as a primary energy source to produce hydrogen and electricity. In this value chain, the first step is the production of nuclear hydrogen from pure water and the extraction of iron ore from mines. Next, iron ore is reduced by nuclear hydrogen [26]. The product of this process is sponge iron. To achieve complete decarbonization of the steel-making supply chain, sponge iron is then refined in electric arc furnaces with the addition of scrap iron. The furnaces can be powered by the electricity produced by the nuclear power plant. Because of the reliability of supply and the large amounts of hydrogen that can be produced, nuclear energy would be ideal for supporting hydrogen production for big consumption centres, and the steel industry can benefit greatly from this potential. The value creation for the different stakeholders involved is described in Table 3.

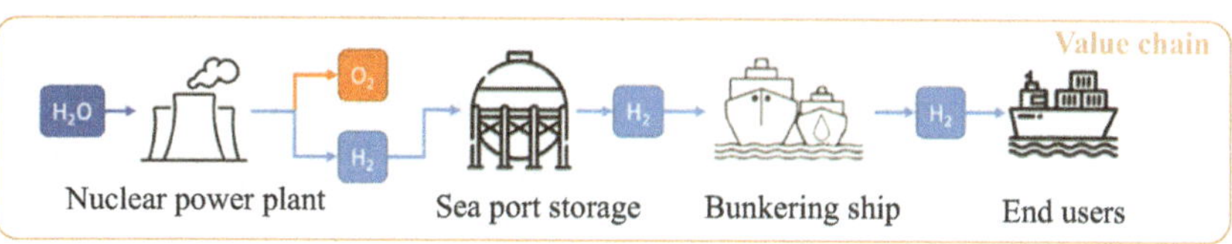

FIG. 8. Nuclear hydrogen shipping value chain.

TABLE 2. VALUE CREATION FOR DIFFERENT STAKEHOLDERS INVOLVED IN THE CHAIN OF NUCLEAR HYDROGEN FOR THE MARITIME TRANSPORT MARKET

Component	Stakeholder	Value
Nuclear power plant	Nuclear power plant owner/ operator	New investments, market growth, diversification, possibility of cogeneration
	Nuclear regulator	Growing interest, new standards
Seaport storage	Seaport operators	Alternative fuel availability, new market segment served, security of supply, contribution to climate commitments, possible diversification strategies
	Seaport constructors	Market growth, development of new competencies
	Environmental groups and non-governmental organizations	Lower environmental impact, CO_2 emission reduction
Bunkering ship	Bunker supply contractors	Security of supply, price stability, service guarantee
	Bunkering vessels constructors	Development of new competencies, possibility of diversification, market growth
End users	Ship owners	Lock-in advantages, decreased emissions
	Ship operators	Decreased uncertainty, security of supply, price stability, contribution to climate commitments

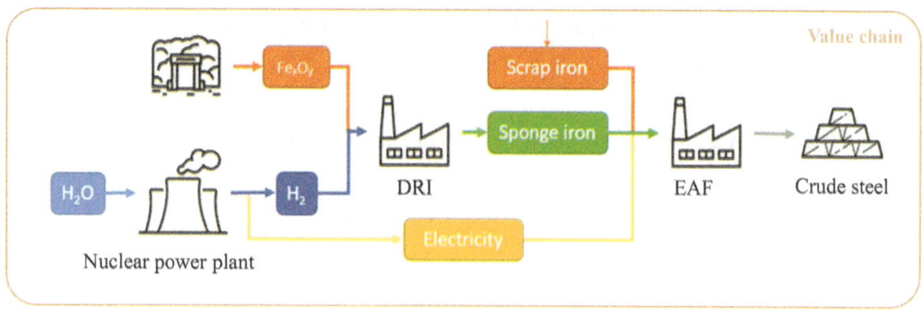

FIG. 9. Nuclear hydrogen steel-making value chain.

TABLE 3. VALUE CREATION FOR DIFFERENT STAKEHOLDERS INVOLVED IN THE CHAIN OF NUCLEAR HYDROGEN FOR THE STEEL MANUFACTURING MARKET

Component	Stakeholder	Value
Nuclear power plant	Nuclear power plant owner/ operator	New investments, market growth, diversification, possibility of cogeneration
	Nuclear regulator	Growing interest, new standards
DRI/electric arc furnaces	DRI operators	Market growth, diversification, improved reputation, price stability, security of long term supply
	DRI facility constructors	New investments, R&D projects, new competences
	Operators of electric arc furnaces	Market growth, diversification, improved reputation
	Environmental groups and non-governmental organizations	Lower environmental impact, CO_2 emission reduction
End users	Manufacturers of steel products	Low carbon steel products, contribution to climate commitments, reduced carbon taxes
	Constructors using steel	Higher acceptability, competitive advantage, differentiation strategies opportunity

3.2. COST CONSIDERATIONS FOR NUCLEAR HYDROGEN PRODUCTION

3.2.1. Hydrogen production cost — drivers and estimates

The main drivers of clean hydrogen costs are:

- Capacity factor of the energy system that supplies the hydrogen production process;
- Capital cost of the energy supply for the hydrogen production process;
- Efficiency of the hydrogen production process;
- Capital cost of equipment for the hydrogen production process.

The effect of each of the factors in the case of hydrogen production through electrolysis is illustrated in Figs 10–12. In general, for the same electrolyser efficiency and capacity factor, the cost of produced hydrogen would increase in line with system and electrolyser capital expenditures (CAPEX). The hydrogen cost is lower when the hydrogen is produced through high temperature electrolysis (HTE) than through low temperature electrolysis (LTE). The details on the assumptions can be found in Ref. [17] .

The cost projections up to 2030 and actual costs for the period 2019–2024 for hydrogen produced through electrolysis using solar PV, wind, nuclear and geothermal energies are indicated in Fig. 13 [17].

A more detailed sensitivity analysis of hydrogen production costs using nuclear energy was performed by the IAEA in 2023. Different levels of electricity generation cost and electrolyser cost were analysed. The results are indicated in Fig. 14, with the following assumptions:

- Discount rate of 7%;
- Operating expenditures of the electrolyser: US $10/kW + 1% of capital costs;
- Lifetime: 30 years;
- Construction time: 1 year;
- Efficiency of the electrolyser of 70%.

3.2.2. Competitiveness of nuclear hydrogen production

The costs of producing hydrogen from the different energy sources, the modes of transportation and the type of energy to be used are all factors that directly impact the relative competitiveness and the strengths and weaknesses of each hydrogen production technology. The value ranges of Figs 10–14 provide

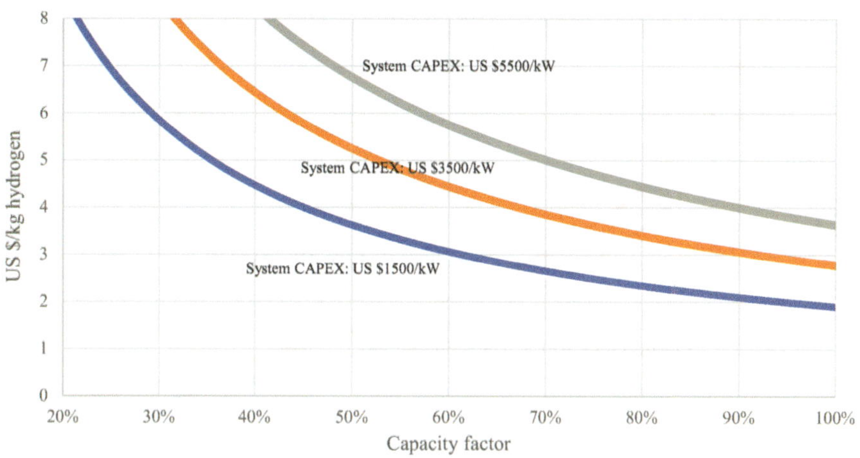

FIG. 10. Influence of the energy system CAPEX on the hydrogen production cost, for an electrolyser efficiency of 0.64.

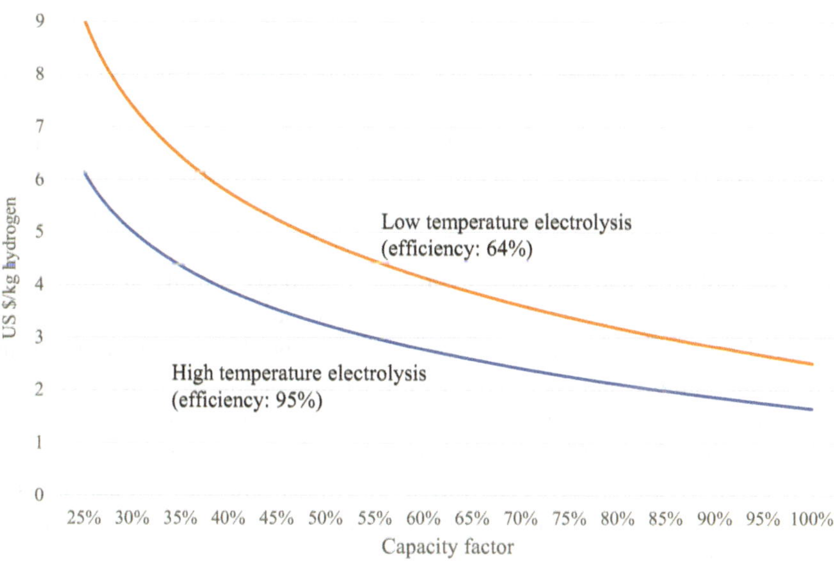

FIG. 11. Influence of the electrolyser efficiency on the hydrogen production cost.

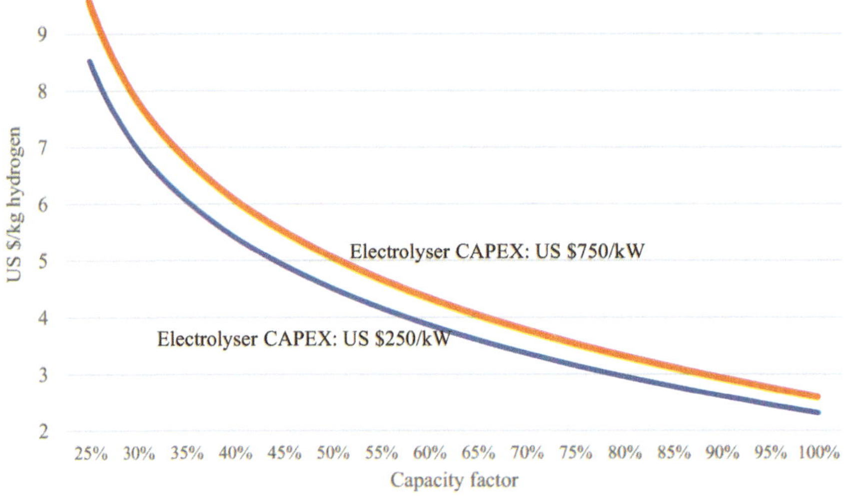

FIG. 12. Influence of the electrolyser CAPEX on the hydrogen production cost.

an example of cost variation. However, the competitiveness depends on a variety of factors, including the quality of the energy source used, its location relative to the end user, the maturity of the process involved and the geography of the location, and the competitiveness can change significantly over time. Ultimately, the competitiveness of different energy sources has to be looked at from the point of view of the end user and include all related costs.

The policies and national hydrogen roadmaps already announced by several countries show that the commercial low carbon hydrogen space could be shared by multiple production options including steam methane reforming with carbon capture and renewable hydrogen production by water electrolysis or steam electrolysis (where a low carbon heat source is also available). In nuclear equipped countries, a nuclear hydrogen supply programme would further diversify this space, since nuclear electricity and heat can support these production technologies. Other upcoming process alternatives that can make use of any low carbon heat and electricity sources include methane pyrolysis and thermochemical cycles.

The availability of multiple options also creates market competition in the hydrogen production sector, which implies that strategic choices will have to be made by countries in this regard. Some examples are cited to illustrate some of the trade-offs involved in this aspect:

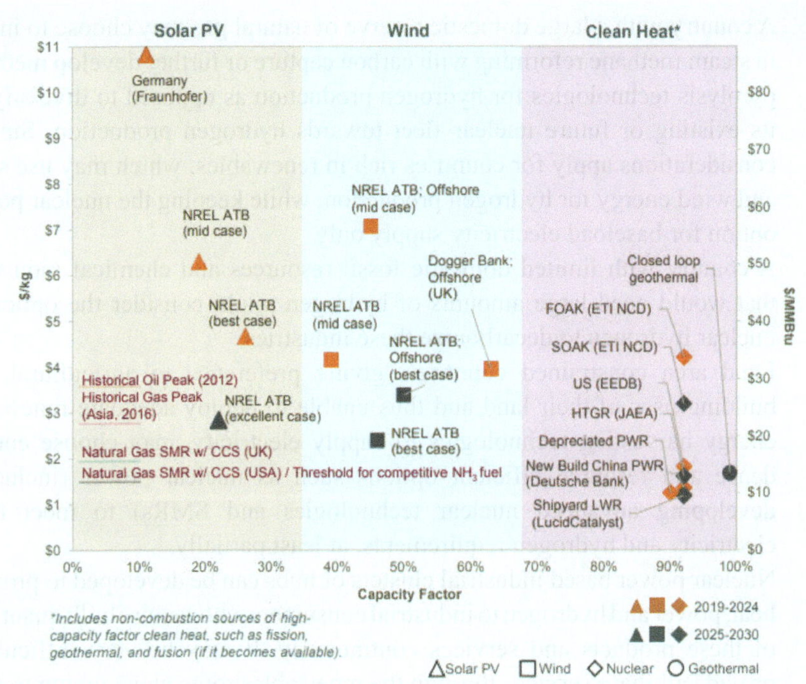

FIG. 13. Projections of hydrogen production cost by 2030 and actual costs for the period 2019–2024 (reproduced courtesy of LucidCatalyst). EEDB — Energy Economic Data Base Program of Oak Ridge National Laboratory; ETI — Energy Technology Institute; FOAK — first-of-a-kind; LWR — light water reactor; MMBtu — million British thermal units (a unit of energy commonly used to measure the heat content of fuels, particularly natural gas); NCD — nuclear cost drivers; NREL ATB — National Renewable Energy Laboratory's Annual Technology Baseline; PWR — pressurized water reactor; SOAK — second-of-a-kind.

Cost of hydrogen generation – nuclear power																	
	Cost of electrolyser (US $/kW)																
Electricity generation costs (US $/MWh)	**200**	**250**	**300**	**350**	**400**	**450**	**500**	**550**	**600**	**650**	**700**	**750**	**800**	**850**	**900**	**950**	**1000**
30	1.6	1.6	1.7	1.7	1.7	1.7	1.8	1.8	1.8	1.9	1.9	1.9	1.9	2.0	2.0	2.0	2.1
35	1.8	1.9	1.9	1.9	2.0	2.0	2.0	2.0	2.1	2.1	2.1	2.1	2.2	2.2	2.2	2.3	2.3
40	2.1	2.1	2.1	2.2	2.2	2.2	2.2	2.3	2.3	2.3	2.4	2.4	2.4	2.4	2.5	2.5	2.5
45	2.3	2.3	2.4	2.4	2.4	2.5	2.5	2.5	2.5	2.6	2.6	2.6	2.7	2.7	2.7	2.7	2.8
50	2.6	2.6	2.6	2.6	2.7	2.7	2.7	2.7	2.8	2.8	2.8	2.9	2.9	2.9	2.9	3.0	3.0
55	2.8	2.8	2.8	2.9	2.9	2.9	3.0	3.0	3.0	3.0	3.1	3.1	3.1	3.2	3.2	3.2	3.2
60	3.0	3.1	3.1	3.1	3.1	3.2	3.2	3.2	3.3	3.3	3.3	3.3	3.4	3.4	3.4	3.4	3.5
65	3.3	3.3	3.3	3.3	3.4	3.4	3.4	3.5	3.5	3.5	3.5	3.6	3.6	3.6	3.7	3.7	3.7
70	3.5	3.5	3.6	3.6	3.6	3.6	3.7	3.7	3.7	3.8	3.8	3.8	3.8	3.9	3.9	3.9	4.0
75	3.7	3.8	3.8	3.8	3.9	3.9	3.9	3.9	4.0	4.0	4.0	4.1	4.1	4.1	4.1	4.2	4.2
80	4.0	4.0	4.0	4.1	4.1	4.1	4.1	4.2	4.2	4.2	4.3	4.3	4.3	4.3	4.4	4.4	4.4
85	4.2	4.2	4.3	4.3	4.3	4.4	4.4	4.4	4.4	4.5	4.5	4.5	4.6	4.6	4.6	4.6	4.7
90	4.5	4.5	4.5	4.5	4.6	4.6	4.6	4.7	4.7	4.7	4.7	4.8	4.8	4.8	4.8	4.9	4.9
95	4.7	4.7	4.7	4.8	4.8	4.8	4.9	4.9	4.9	4.9	5.0	5.0	5.0	5.1	5.1	5.1	5.1
100	4.9	5.0	5.0	5.0	5.0	5.1	5.1	5.1	5.2	5.2	5.2	5.2	5.3	5.3	5.3	5.4	5.4

FIG. 14. Results of the sensitivity analysis of hydrogen production cost performed by the IAEA; the blue bars indicate the region of target electrolyser and generation cost.

— A country with a large domestic reserve of natural gas may choose to invest in steam methane reforming with carbon capture or further develop methane pyrolysis technologies for hydrogen production as opposed to diversifying its existing or future nuclear fleet towards hydrogen production. Similar considerations apply for countries rich in renewables, which may use solar and wind energy for hydrogen production, while keeping the nuclear power option for baseload electricity supply only.
— A country with limited domestic fossil resources and chemical industries that would need large amounts of hydrogen might consider the option of nuclear hydrogen to decarbonize these industries.
— Land area constrained countries, giving preference to agricultural and building uses of their land and thus unable to deploy adequate renewable energy harvesting technologies to supply electricity, may choose energy dense and resource efficient options such as nuclear power (including developing advanced nuclear technologies and SMRs) to meet their electricity and hydrogen requirements, at least partially.
— Nuclear power based industrial clusters or hubs can be developed to provide heat, power and hydrogen to industrial consumers who require bulk quantities of these products and services continuously. It may be very difficult to provide all these services through the renewables route alone owing to land area limitations, the intermittency and variability of renewables, and the huge need for energy/material storage.
— Countries with large oil and gas reserves may find it opportune to invest in nuclear energy to decarbonize and diversify their energy mix while simultaneously ensuring that fossil fuel resources continue to be available for other applications such as chemicals synthesis, for which alternative feedstock is still not available. These countries may also develop nuclear hydrogen programmes as a feasible alternative to meet domestic requirements.
— Countries with large nuclear power plants in coastal locations may look to produce nuclear hydrogen or hydrogen derivatives specifically for the decarbonization of the maritime sector by supplying to fuel bunkers located at nearby ports.

The final hydrogen production technology mix that a country adopts will therefore not be a standalone decision; it will be determined by holistically considering various factors unique to the country, including resource adequacy and the country's need for diversification, expected hydrogen demand and demand growth in end use sectors, electricity production processes, availability of hydrogen production technology, supply security of materials and components for hydrogen production, economic cost benefits of domestic production (vis-à-vis imports) and the overall decarbonization pathway adopted by the country.

3.3. ECONOMIC PERSPECTIVES OF COLLOCATION AND COUPLING OF THE HYDROGEN PRODUCTION PLANT WITH THE NUCLEAR POWER PLANT

Economic and viability factors of collocating and coupling of a hydrogen production plant with a nuclear power plant are discussed in this section. For the hydrogen production plant, LTE methods, such as alkaline and polymer electrolyte membrane (PEM), require only electricity to convert water to hydrogen, while HTE methods require electricity and heat to provide energy to split water. LTE and HTE are considered leading technologies for hydrogen production using a nuclear power plant (see Section 4 for more technical details).

3.3.1. Collocation

Collocation within the context of this publication is defined as the location and siting of a hydrogen production plant and transient daily storage of produced hydrogen inside the fence of a nuclear power plant. Inside the fence of a nuclear power plant are separated areas or zones, as shown in Fig. 15. In this configuration, the connections between the hydrogen production plant and the nuclear power plant are intrinsically interlinked, shared services are possible and transportation/transmissions costs are reduced.

3.3.2. Coupling

In the context of this publication, coupling is defined as the locating or siting of hydrogen production and transient daily storage of produced hydrogen outside the fence of a nuclear power plant. The distance away from the fence depends on the accessibility of energy from the power plant without a significant reduction in efficiency. Hence, a coupled facility is considered a location outside both active and non-active areas of the nuclear power plant. The layout shown in Fig. 16 and considered for discussion in this publication locates the hydrogen production plant outside the site fence of the nuclear power plant, at a distance away from the licensing boundary of the nuclear power plant. Again, it is assumed there are economic and technology related attributes to justify such coupling to supply electricity and heat.

There are various economic factors that influence the collocation and/or coupling of a nuclear power plant with a hydrogen production plant. Among

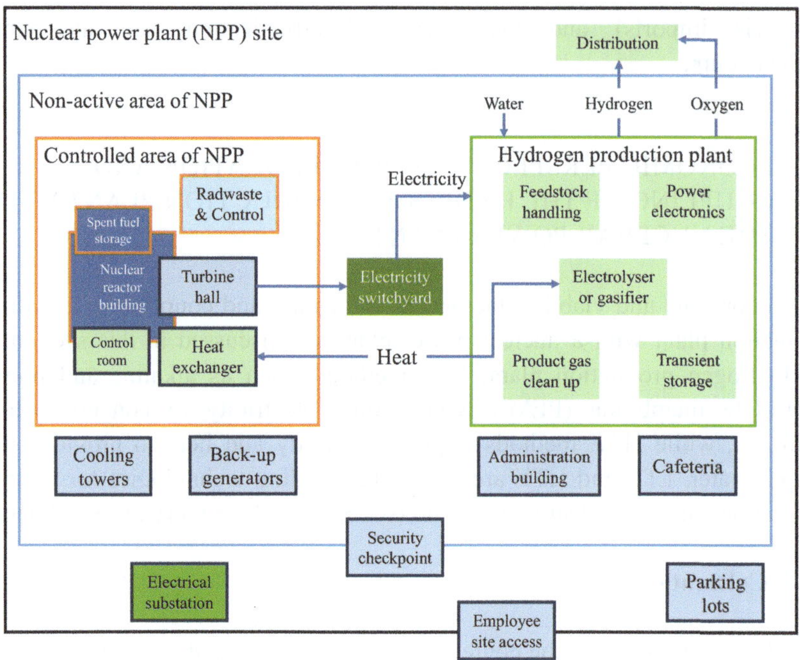

FIG. 15. Collocation of a hydrogen production plant with a nuclear power plant. The illustration is solely a representation of a general scenario, which may vary for specific locations. It is intended to illustrate a point of collocation in the broadest terms. Note: The components of the hydrogen production plant are shown in green; the components of the nuclear power plant are shown in blue.

them are the energy conversion efficiency from the electricity generated from the nuclear asset to the hydrogen produced by the hydrogen plant; energy availability factors; and siting, capital and operation and maintenance (O&M) costs. Tables 4–8 present some considerations as to how the levelized cost of hydrogen (LCOH) might be affected, depending on whether the hydrogen plant is collocated or coupled with the nuclear power plant, by factors such as:

- The energy conversion efficiency (from nuclear electricity into hydrogen, from nuclear heat into hydrogen) (see Table 4);
- Nuclear energy availability for hydrogen production (see Table 5);
- Siting costs associated with the hydrogen production facility and cost associated with the hydrogen transport and distribution (see Table 6);
- Capital costs of the nuclear power plant and of the nuclear hydrogen production facility (see Table 7);
- Fixed and variable O&M costs (see Table 8).

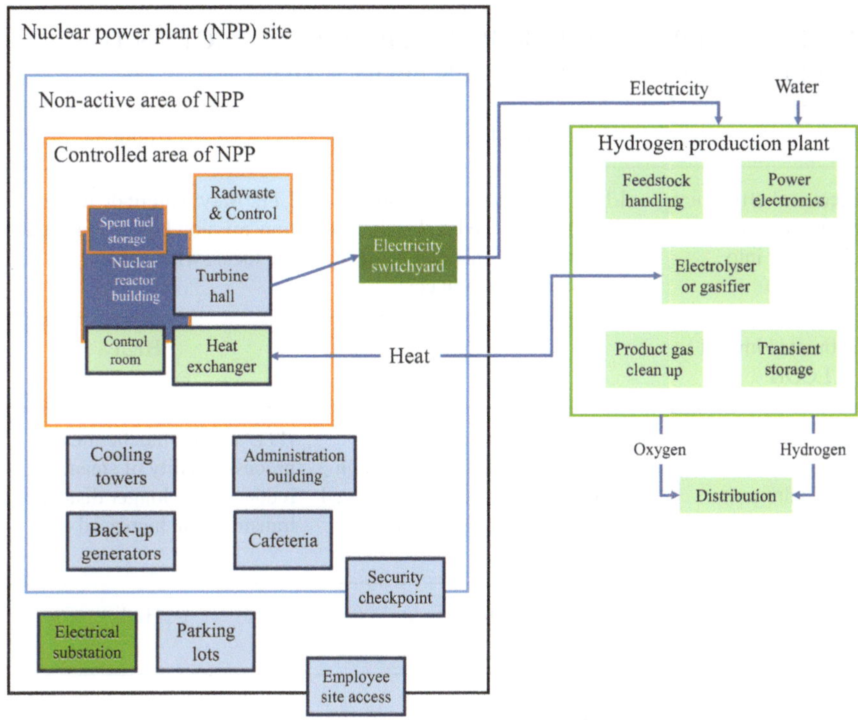

FIG. 16. Coupling of a hydrogen production plant with a nuclear power plant. The illustration is solely a representation of a general scenario, which may vary for specific locations. It is intended to illustrate a point of coupling in the broadest terms. Note: The components of the hydrogen production plant are shown in green; the components of the nuclear power plant are shown in blue.

Table 4 presents the energy conversion efficiency impact on the levelized cost of hydrogen production. The amount of heat that has to be discharged for the electricity output is determined by a plant's thermal efficiency, which is the proportion of internal heat that becomes electrical output. Electrical efficiency shows the difference between the gross electrical output of the plant and the amount of electricity used for LTE. While electrical efficiency has a high impact on LCOH in both collocation and coupling options, the thermal efficiency has a medium impact on LCOH for collocation and a high impact for coupling.

TABLE 4. ENERGY CONVERSION EFFICIENCY IMPACT ON THE LEVELIZED COST OF HYDROGEN PRODUCTION

	Collocation	Coupling
Electrical efficiency (from nuclear electricity into hydrogen)	LTE: Ability to connect power directly from the local switchyard to reduce transmission loss and improve production cost	LTE: May require dedicated transmission lines connected to the local switchyard
Estimated impact on LCOH	High	High
Thermal efficiency (from nuclear heat into hydrogen)	HTE and thermochemical process: less heat loss and high quality of steam can be achieved owing to short distance for transmission	HTE and thermochemical process: quality of steam is reduced with longer distance for transmission, increased cost
Estimated impact on LCOH	Medium	High

Table 5 shows possible impacts of collocation and coupling on energy availability for existing and future nuclear power plants. Hydrogen production may affect the baseload of the nuclear power plant, flexibility of the electricity production and, thus, the production costs. These potential aspects have to be carefully considered when deploying both LTE and HTE hydrogen production. While collocation/coupling with the existing nuclear power plant has a high impact on LCOH, collocation/coupling with the future nuclear power plant has a medium impact on LCOH.

TABLE 5. ENERGY AVAILABILITY IMPACT ON THE LEVELIZED COST OF HYDROGEN PRODUCTION

	Collocation	Coupling
Existing nuclear power plant	— The off-peak or excess electricity can be used for hydrogen production. This can essentially impact hydrogen production costs — The hydrogen production option has to bring added value for the utilities in comparison with ensuring the electricity supply for other consumers — The collocation is unlikely to have more advantages for LTE than for HTE, while for HTE and thermochemical processes, it brings the opportunity to access steam and heat	Same as collocation
Estimated impact on LCOH	High	High
Future nuclear power plant	— Flexible electricity and heat supply, as well as ability to operate outside baseloads to improve cost of hydrogen production — Operational strategy, design and layout can be optimized for cost reduction — No more advantages for LTE than for HTE, while for HTE and thermochemical processes, it brings the opportunity to access steam and heat	Same as collocation
Estimated impact on LCOH	Medium	Medium

Siting costs can vary significantly when collocating or coupling a hydrogen production plant with the nuclear power plant. For collocation, most siting costs are already included in the nuclear power plant costs, but for coupling, additional infrastructure and siting costs have to be considered for the hydrogen production plant. Siting costs have high impact on LCOH in case of collocating, and medium impact in case of coupling, as shown in Table 6.

TABLE 6. SITING COST IMPACT ON THE LEVELIZED COST OF HYDROGEN PRODUCTION

	Collocation	Coupling
Costs	— Site preparation, arrangement a safe distance from the hydrogen production plant; additional safety features for the hydrogen production plant — No land cost — Potential savings from sharing common facilities	— Site selection and site preparation, land cost — Additional costs for hydrogen transportation (short, medium or long distance) — No costs added to additional safety features — No savings from sharing common facilities
Steam transmission and hydrogen distribution	— For HTE, lower steam transmission cost due to shorter transportation distance and pipeline — If distributing hydrogen by truck, the costs are the same as for coupling	— For HTE, higher steam transmission cost due to longer transportation distance and pipeline — Potential cost savings from injecting hydrogen into a natural gas pipeline, which may not be accessible for collocation
Estimated impact on LCOH	Medium	High

Table 7 shows the impact of capital costs for a nuclear power plant and a hydrogen production plant.

TABLE 7. CAPITAL COST IMPACT ON THE LEVELIZED COST OF HYDROGEN PRODUCTION

	Collocation	Coupling
Nuclear power plant	— For existing nuclear power plants, the capital cost might have already been paid off. Plant modification is required for coupling with HTE and thermochemical cycles — For future nuclear power plants, cost is part of design and construction — There is also a licensing associated cost (both for existing and future nuclear power plants)	Same as collocation
Estimated impact on LCOH	Medium	Medium
Hydrogen production facility	— Alkaline electrolysis has the lowest initial cost and no collocation benefit — PEM has a lower investment cost than HTE but higher than LTE and there is no collocation benefit — HTE and thermochemical cycles have higher efficiencies than LTE but higher investment cost	Similar capital investment as for collocation
Estimated impact on LCOH	High	High

O&M costs can be split into fixed and variable costs. Fixed costs represent the costs of power plant operations and maintenance that are happening during the operation and shutdown of a nuclear power plant (e.g. plant preventative maintenance, inspections, monitoring). The impact of O&M costs on the LCOH is presented in Table 8.

TABLE 8. O&M COST IMPACT ON THE LEVELIZED COST OF HYDROGEN PRODUCTION

	Collocation	Coupling
Fixed O&M costs	— Slight increase to nuclear power plant overhead costs due to the additional hydrogen production plant and liabilities — Lower O&M cost due to shorter distance from the nuclear power plant to the hydrogen production plant — Hydrogen production plant is subject to nuclear liability and property insurance, higher O&M costs — Insurance provider may determine that it cannot insure the nuclear power plant with the collocated hydrogen production plant — Ability to share common facilities, saving O&M costs	— Lower O&M costs due to labour and to parts — For HTE, slightly higher O&M for the steam system due to long distance from nuclear power plant to hydrogen production plant — The hydrogen production plant is not subject to nuclear liability and property insurance, lower O&M costs — No concerns with nuclear liability and property insurance
Estimated impact on LCOH	Low	Low
Variable O&M costs	— Electricity price — Potentially long refuelling cycles for some SMR designs can be translated into reduced handling and transport requirements of nuclear materials and cost — Costs to transport feedstock across boundary of nuclear power plant — Carbon pricing in different jurisdictions	— Electricity price — Costs to transport feedstock — Carbon pricing in different jurisdictions
Estimated impact on LCOH	Low to medium, except electricity price that has a high impact	Low, except electricity price that has a high impact

3.4. POSSIBLE BUSINESS MODELS

For a nuclear hydrogen project, the following configurations can be considered, depending on the location of the nuclear power plant and the end users of the hydrogen produced. These configurations consider that a nuclear power plant owns a hydrogen production plant in its vicinity (either collocated or coupled) and has the right to sell the hydrogen produced.

- The 'close to the user' configuration might be considered when the nuclear power plant produces hydrogen for only one consumer (e.g. for a large scale steel-making factory). This hydrogen can then be sold directly and only to this consumer, at a fixed price. This might also be the case if a large industrial consumer owns the nuclear power plant.
- The 'close to users' configuration might be considered for the case where the nuclear power plant produces hydrogen that is delivered within one specific region (e.g. within a radius of 200 km from the nuclear power plant). This configuration might not need pipeline infrastructure, as it might be cheaper to transport the hydrogen by truck. The selling prices may be established locally, as hydrogen is produced within the region and there is no need to involve a hydrogen distribution company. Distribution infrastructure such as compressors and intermediate storage may still be needed.
- The 'nuclear power plant is located a long distance from the hydrogen end user' configuration might be considered when the nuclear hydrogen is distributed by regional companies that buy the hydrogen and sell it to local industries in their regions or that keep the hydrogen in storage and sell it on demand. In this scenario, the deployment of hydrogen hubs might be considered, based on operating CO_2 hubs and clusters, as described in Ref. [69]. Hydrogen hubs may collect hydrogen from many sources (e.g. nuclear, renewable, fossil fuel power plants) and distribute it to single or multiple storage locations. Hydrogen clusters may collect individual hydrogen sources or storage sites within the selected region.

3.4.1. Hydrogen valleys and hydrogen hubs

A hydrogen valley or industrial hub is a practical realization of the hydrogen ecosystem concept. It consists of one or more hydrogen producers and more than one consumer linked together by a hydrogen distribution and storage network, all located in close geographical proximity [70]. This approach helps create a ready market for hydrogen and assists in the development of a business case for a hydrogen application for specific sectors. The approach has been widely discussed in Europe and the United States of America (USA), and hydrogen valleys are also under development in Latin America, India and Africa. While most hydrogen hubs have been envisaged for renewable hydrogen production, the concept may be extended to nuclear hydrogen production projects as well. Various configurations connecting producers and consumers can be conceptualized, giving rise to different business models.

3.4.2. Business models with small modular reactors

It is expected that shifting from traditional on-site nuclear construction projects to factory built, modular components will ensure lower capital and operating costs, shorter construction times and lower financial risks for the deployment of nuclear reactors. Different deployment models can be used to achieve this. For example, one model is to include a fully integrated facility that manufactures modular components that are then installed and operated on the same site for efficient, low cost and large scale production of hydrogen and clean synthetic fuels. This could enable rapid, affordable decarbonization of carbon intensive sectors by bringing in innovations in the form of design for manufacture and assembly, standardization, modularization, collocation and cogeneration, such as in the case of the gigafactory concept (see Section 3.4.2.1). Another deployment model is the shipyard manufactured production platform. This is based on the floating production, storage and offloading facilities of existing large oil and gas industry vessels. Advantages of offshore siting include the possibility to supply multiple products (e.g. power, fuel, fresh water) to large coastal cities without requiring major additional investments in terrestrial infrastructure projects; the possibility to variably serve electrical power production or hydrogen fuel production, complementing solar power; and the elimination of land use challenges and siting issues. Both deployment models offer potential pathways for delivering the necessary hardware at the required global scale.

The cost of producing clean hydrogen and synthetic fuels will be significantly reduced by switching from traditional building to highly productive shipyard manufacturing. As explained below, current shipyard production has a large capacity to supply facilities for the production of hydrogen specifically constructed for this intended use. It may also be expanded and modified to accommodate growing demand.

3.4.2.1. The hydrogen gigafactory model

A hydrogen–synthetic fuel gigafactory is a large, integrated facility that produces clean hydrogen and synthetic fuels efficiently and cost-effectively (e.g. it could produce 50 Mt of ammonia annually). It can be located on former coastal refinery sites and directly connect to existing gas networks. The factory can produce significant amounts of ammonia or synthetic hydrocarbon fuel annually and may also house other facilities that use hydrogen as a feedstock, benefiting from low cost electricity and hydrogen [17].

3.4.2.2. The shipyard manufactured fuel production platform

Shipyards have become highly efficient manufacturing environments, particularly for large scale fabrication. Intense competition and growing demand have led to world-class design, manufacturing and quality assurance programmes. These shipyards are well-suited for creating high quality, cost-effective hydrogen–synthetic fuel production platforms on schedule and in large quantities. Offshore siting of such platforms would facilitate the supply of clean energy and multiple products (such as ammonia, synthetic aviation fuel and desalinated water) to coastal cities without needing major terrestrial infrastructure investments. These platforms complement solar power and avoid land use challenges and siting issues near population centres [17].

3.5. FINANCING OF NUCLEAR HYDROGEN PROJECTS

In general terms, the cost of a hydrogen project would be lower than the cost of a nuclear power plant. There is an incentive to utilize, where possible, the existing nuclear capacities to produce hydrogen, as this can increase the profitability as well as the performance of the nuclear power plant. The possible financing mechanisms and their applicability to nuclear hydrogen projects are presented in this section. A country can adopt one mechanism or a combination of mechanisms through appropriate innovations in financing [71]. Some of the possible mechanisms are the following:

- Adoption through annual national budgets. National governments with an established commitment to developing and/or maintaining a nuclear power programme and a hydrogen strategy can provide funding through annual budget allocation to fully support the deployment, operation and development of the nuclear hydrogen projects, at least in the case of a first of its kind project.
- Ensure revenues from hydrogen production. An established nuclear hydrogen project with clearly identified offtakers and long term purchase contracts can generate adequate revenues to meet a large part of the continued financing needed to sustain and even expand the projects, thereby reducing dependence on external sources of financing. The possible business mechanisms are listed in Section 3.4.

- Establishing public–private partnerships and other forms of cooperative financing models. Nuclear hydrogen projects may be jointly financed, built, owned and operated by the public and/or private sector organizations over 25–30 year time frames. (The 'Mankala' model[2] deployed in the Finnish nuclear industry is an example of this financing arrangement [72, 73].) This works best with power and hydrogen intensive industries or industrial clusters.
- Establishing dedicated finance/impact investing opportunities. Specific financial instruments such as sovereign and corporate green bonds or sustainability bonds may be issued to obtain access to capital markets to raise capital for nuclear hydrogen projects. This approach may be well supported by the adoption of a sustainable finance taxonomy that covers nuclear power and cogeneration projects.
- Financing by development finance institutions. Multilateral banks could issue long term, low cost loans to finance the nuclear hydrogen projects as long term clean energy/industry infrastructure projects, particularly in embarking countries.
- Adopting regulated asset base models. Full investment, operation and ownership of the hydrogen plant (and possibly the hydrogen distribution infrastructure as well) can be retained by the private sector project developer, who can earn revenues from the hydrogen provided to end users. This may help spread out the risk over a large user base and it may be possible to reduce financial risk perception of the project.
- Establishing project financing models. An independent legal entity may be created to act as the owner of a nuclear hydrogen project and obtain financing from an independent set of financers. All operating expenses for the project are covered by the revenue generated by the project owner. The capital equipment used in the project, such as machinery, technology and infrastructure, serves as collateral. This means that these assets are pledged as security for any loans taken out by the project owner.

3.6. REVENUES FROM A NUCLEAR HYDROGEN PROJECT

Securing revenue — and diversifying its sources — is key to attracting investors, lowering the cost of capital and enabling the development and deployment of nuclear hydrogen projects. These projects can benefit from

[2] In the Mankala financing model, the shareholders (usually from the industry and utilities) purchase electricity from the power plant equal to their shareholding at the cost price. The electricity can then be used by the shareholders, as needed, or sold.

existing revenue frameworks, developed initially for electricity generating power plants, which can be adapted and extended to cover nuclear hydrogen. This section provides an overview of the main frameworks that can be applied. It also suggests a few policy levers (e.g. tax reliefs, subsidies) for consolidating revenues and potentially improving the profit margins of nuclear hydrogen projects.

3.6.1. Revenue frameworks

The main remuneration tools that can be applied to nuclear hydrogen are presented in the following paragraphs. These are existing frameworks that are mainly used to remunerate nuclear electricity producers.

3.6.1.1. Power purchase agreements and feed-in tariffs

A power purchase agreement is a contractual agreement between a buyer (e.g. a hydrogen distribution company) and a seller (e.g. a hydrogen producing company or an independent power producer offering hydrogen as a by-product in addition to electricity). Power purchase agreements are usually associated with large supply, often uncorrelated with the demand level and delivered cheaply, typically from variable renewable energy sources.

The feed-in tariffs can be part of a power purchase agreement in which fixed prices are paid to producers. Feed-in tariffs are usually calculated on the basis of the levelized cost of electricity or hydrogen (in the case of hydrogen projects).

3.6.1.2. Contract for difference and regulated asset base models

A contract for difference is a financial contract that pays the differences in the settlement price between the strike price (the fixed price agreed upon when the contract is created) and the actual (market) price. The contract for difference scheme is the UK Government's main support mechanism for new low carbon electricity generation projects — the Hinkley Point C project, for example [74].

Another framework proposed for nuclear newbuild projects in the United Kingdom is the regulated asset base model, which is designed to meet a utility's revenue requirements based on the capital it deploys to provide a service of public utility (e.g. electricity) [75]. Infrastructure projects offering hydrogen based energy vectors — and seasonal storage services for hydrogen based electricity — can also benefit from such mechanisms [75].

3.6.1.3. Cooperative models

In a cooperative financing model (such as the Mankala model in Finland [72, 73]), a group of investors raise debt and equity for a project and share the risk related to it [76]. Using a cooperative financing model for nuclear hydrogen projects, the end users can also be shareholders, which would allow them to purchase a certain percentage of hydrogen at production cost. The cooperative models are particularly suited for energy and hydrogen intensive industries, for which security of supply and price certainty are of utmost importance.

3.6.2. Revenue-consolidating policy levers

Carbon pricing is an instrument used by governments to incentivize the development of clean hydrogen projects. (Additional instruments include production subsidies, carbon credits and other regulations.) Carbon pricing can take the form of a direct carbon tax or tradable carbon emission rights (referred to as an emissions trading system). By 2022, the share of global GHG emissions covered by carbon pricing instruments reached around 23% [77] in the top ten ranked GHG emitters (nine countries and the European Union), the majority of which use nuclear energy [78].

A forecast of the impact of carbon pricing on hydrogen production costs in the European Union in 2030 by hydrogen type is shown in Fig. 17, using data from Statista (as of 2021) [79]. The reference year is 2021, the gas price is set at €20 per MWh, the capture rate for fossil based hydrogen with carbon capture

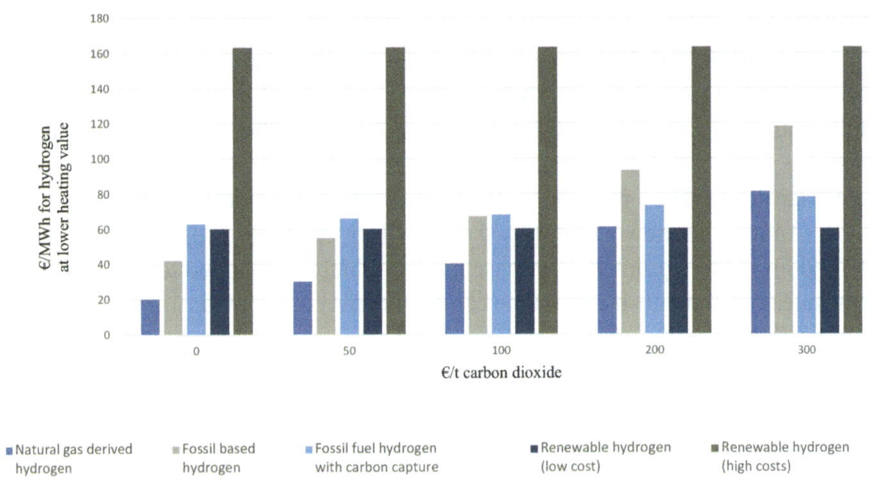

FIG. 17. Forecast impact of carbon pricing on hydrogen production costs in the European Union in 2030, by hydrogen type [79].

is around 75%, and the production cost of renewable hydrogen is between €60 per MWh and €163 per MWh.

Figure 17 indicates that for the cheapest renewable hydrogen to be competitive with fossil based hydrogen in 2030, carbon pricing needs to climb to about €100 per tonne of CO_2. A similar argumentation would apply to nuclear hydrogen. It also shows that natural gas derived hydrogen would remain cheaper than any hydrogen production method if carbon prices do not exceed €100 per tonne of CO_2. However, this happened before the gas price spike after 2021.

It is widely recognized that carbon pricing alone is unlikely to be sufficient, and therefore carbon pricing needs to be part of an integrated package of complementary policies [8] that could include revenue frameworks as described in the previous section. Inclusion in the taxonomy legislation might also be considered. For example, the provision in the Gas Directive of the European Union, proposed at the end of 2022 [80, 81], allows low carbon hydrogen to be counted towards the achievement of decarbonization targets for industry and transport under the European Union's Renewable Energy Directive [81].

3.7. BARRIERS AND ENABLERS FOR THE NUCLEAR INDUSTRY IN THE HYDROGEN MARKET

3.7.1. Impact of various forces on nuclear hydrogen production

Porter's Five Forces framework [82] is a method adopted to examine the industry in which an organization operates and the business pressures that derive from the operating environment. The framework is based on the coexistence of internal and external forces. Internal forces are determined by the rivalry among existing competitors. The higher the existing rivalry, the lower the profitability of the sector and its attractiveness. External forces are determined by the bargaining power of suppliers, the bargaining power of buyers, the threat of substitute products and the possible entry of new competitors. In this publication, the analysis focuses on the hydrogen market and the potential role that nuclear hydrogen plays as a 'new entrant'. Porter's framework considers the coexistence of five industry specific internal and external forces (see Fig. 18):

— Internal forces are due to the internal rivalry among the technologies currently adopted to produce hydrogen (e.g. steam reforming, coal gasification, electrolysis). From this perspective, Section 2 highlighted the current massive adoption of fossil fuels for hydrogen production at low production costs.

- External forces are characterized by the bargaining powers of suppliers and buyers, and the threats of substitutes and new entrants:
  - The bargaining power of suppliers deals with the number of suppliers available, the possibility of choosing different suppliers and the contractual relationships that can be established.
  - The bargaining power of buyers is strictly connected with the switching costs that buyers have to undertake and the range of available choices for the buyers.
  - The threat of substitutes is represented by the existence of products on the market that buyers consider interchangeable with hydrogen. Examples of hydrogen substitutes include batteries, biofuels and fossil fuels with CCS.
  - The threat of new entrants is represented by the possibility of penetrating the hydrogen market with different technologies. The threat level is characterized by high investments, critical expertise and legal requirements.

Considering nuclear hydrogen as a potential new entrant in the market, Porter's Five Forces framework can clearly illustrate the competitive advantage of this technology over existing technologies. Therefore, it is necessary to investigate the advantage that nuclear hydrogen has concerning the bargaining power of suppliers and buyers, the threat of substitutes and the internal market competition.

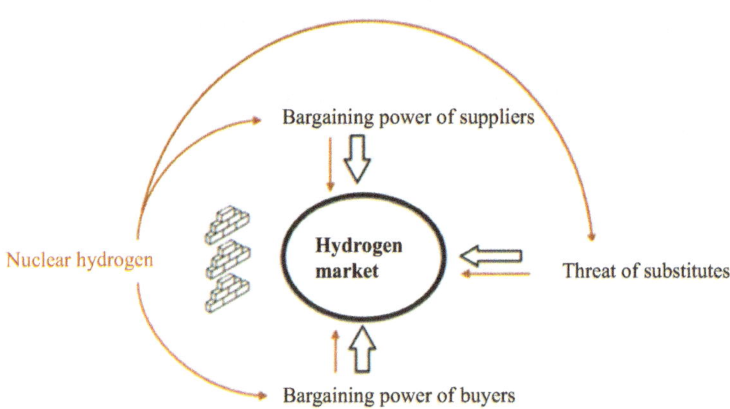

FIG. 18. Porter's Five Forces applied to the nuclear hydrogen market.

— The bargaining power of suppliers: Nuclear hydrogen is characterized by high security of supply, production that can be scheduled and cogeneration possibilities, therefore increasing flexibility. The dependence on suppliers will be lower compared with existing fossil fuel technologies.
— The bargaining power of buyers: Nuclear hydrogen would allow for a stable low price, making alternative production methods less competitive. However, nuclear hydrogen requires high capital investments and a long deployment time, increasing the need to lock in buyers.
— The threat of substitutes: Substitute products and technologies would exist as for existing technologies. However, buyers could be locked into long term contracts, increasing eventual barriers to exit (e.g. switching costs).
— Internal competition: Nuclear hydrogen has advantages such as high availability and low carbon production. However, other internal competitors, such as renewable energy and electrolysers, are economically competitive, putting nuclear hydrogen in a challenging position.

Regarding barriers, nuclear hydrogen production would require a long deployment time, and high CAPEX investments in the case of new nuclear builds. In addition, there is also the cost of establishing the regulations required for coupling a nuclear power plant with a hydrogen production facility. In the case of already existing nuclear power plants, the nuclear hydrogen project would need a shorter deployment time to construct the hydrogen production facility and establish whether the business case is favourable. Despite nuclear hydrogen presenting several advantages when compared with existing technologies, to penetrate the market these barriers need to be overcome. To sum up, all the strengths and weaknesses deriving from the technology and the opportunities and threats deriving from the market are presented in a strengths, weaknesses, opportunities and threats (SWOT) analysis.

3.7.2. SWOT analysis

Similar to Porter's Five Forces framework, SWOT analysis is commonly performed to describe the internal and external contexts of an organization. Figure 19 presents a SWOT analysis describing the strengths and weaknesses of nuclear hydrogen projects and the opportunities and threats that can arise. As mentioned, nuclear hydrogen can guarantee a stable and secure supply at a stable and low price, has low carbon emissions and its production can be scheduled. At the same time, however, it requires a long deployment time and high capital investments. Considering the external context of nuclear hydrogen, several opportunities can be identified in being the 'first mover' (a company or entity that is the first to enter the nuclear hydrogen market), in increasing the range of

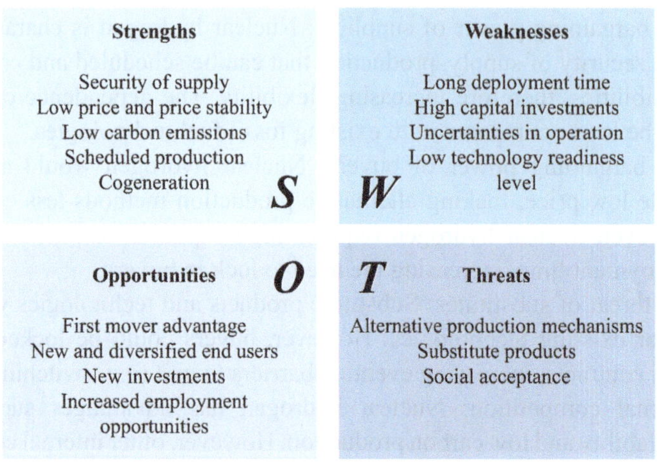

FIG. 19. SWOT analysis for a nuclear hydrogen project.

end users and in attracting new investments. Alternative production mechanisms and substitute products can represent significant threats for nuclear hydrogen.

Table 9 provides examples of acceleration measures for strengths and opportunities and some possible means to mitigate the weaknesses and threats.

3.7.3. Enablers

3.7.3.1. Subsidies and tax credits

In recent years, interest in hydrogen has grown rapidly in many countries, and several countries have developed comprehensive hydrogen strategies reflecting their energy needs, environmental goals and economic targets. Many of these strategies include subsidies, support for investment and favourable regulation to promote hydrogen through production related procurement and end use technologies. The hydrogen strategy for the climate-neutral Europe mentions that support for hydrogen is needed because low carbon hydrogen will require significant investment in the future and it is currently not cost competitive with fossil fuel based hydrogen [83]. On that basis, in September 2022 the European Commission announced up to €5.2 billion in public financial support for 35 low carbon hydrogen related projects [84].

In order to simultaneously stimulate demand and supply in the hydrogen market, government support should be provided not only for individual projects, but also for developing hydrogen clusters that integrate hydrogen production, storage, transportation and final consumption. In the USA, the Infrastructure

TABLE 9. EXAMPLES OF MEASURES TO ACCELERATE THE STRENGTHS AND OPPORTUNITIES AND MITIGATE THE WEAKNESSES AND THREATS

	Pros	Acceleration measures
Internal (Strengths)	Security of supply	Leverage long term contracts Diversify supply of critical components (e.g. fuel)
	Low price and price stability	Fixed price contracts Adoption of derivatives financial instruments
	Low carbon emissions	Leveraging on positive reputation Comparation with other production mechanisms
	Can be scheduled	Diversification strategies (e.g. switching to electricity production for the grid)
External (Opportunities)	First mover advantage	Early investments in R&D, involvement of new actors, development of patents
	New and diversified end users	Involvement of primary and secondary stakeholders, create a well defined network
	New investments	Reducing perceived risks
	Cons	**Mitigation measures**
Internal (Weaknesses)	Long deployment time[a]	Standardization, modularization, early start for investments
	High capital investments	Careful planning of financing, balancing of portfolio, policy support to reduce the perceived risk
External (Threats)	Alternative production methods	R&D investments, early acquisitions, lock-in contracts, proximity to large consumption centres (to reduce transportation cost)
	Substitute products	Creation of high barriers to exit, long term lock-in contracts

[a] Years if existing nuclear infrastructure; >10 years if nuclear infrastructure is to be built.

Investment and Jobs Act [85], signed into law in November 2021, included provisions for the development of clean hydrogen hubs, allocating US $8 billion to support the creation of at least four regional clean hydrogen hubs across the country. Hydrogen hubs are expected to create a network of hydrogen producers, consumers and regional connecting infrastructure to accelerate the use of low carbon hydrogen produced from renewable and nuclear energy sources [86].

While most of these support activities focus on hydrogen production from renewable energy sources, the USA and the UK also provide subsidies for demonstration projects of hydrogen production using nuclear electricity [87, 88]. Several nuclear hydrogen projects are under way in the USA [89]. For example, a project in Arizona will receive funding of up to US $20 million [90]. The goal of this demonstration project is to use the stored hydrogen produced with electricity from a nuclear power plant (Palo Verde Nuclear Generating Station) to deliver approximately 200 MWh of electricity during times of high demand, and may also be used to make chemicals and other fuels. The hydrogen will contribute to a stable supply of electricity in the face of the increasingly variable renewable energy. In the UK, a feasibility study on hydrogen production at nuclear power plants, funded by the government, showed that the technical and regulatory challenges could be cleared [91]. Several projects based on this study are under consideration, including hydrogen production with a high temperature electrolyser that can use both electricity and heat from a nuclear power plant.

The Inflation Reduction Act, enacted and adopted in the USA in 2022, set a subsidies mechanism linked to the carbon intensity of the hydrogen produced [92]: a US $3 subsidy for each kilogram of hydrogen produced if less than 450 g of CO_2 per kg of H_2 is emitted [93]. Thus, hydrogen produced by nuclear energy is eligible for the US $3 per kg subsidy. The act aims to accelerate the deployment of clean technologies in the USA while helping to structure a market, securing the demand for these technologies, and making the investments in clean technologies more attractive and beneficial [94].

In February 2023, the European Commission issued the Green Deal Industrial Plan, which aimed to prevent the outflow of investments in factories or job creation in the emerging green technologies market from the European Union to the USA [95]. For hydrogen, the Inflation Reduction Act [92] set a target of 1:1:1, meaning a cost of US $1 for 1 kg of hydrogen to be achieved in one decade. The establishment of the subsidies mechanism removed roadblocks for hydrogen project developers in the early stage of this emerging market.

Although the European Union did not clearly specify whether nuclear energy could be used for hydrogen production, low carbon hydrogen is included among other decarbonizing technologies in the Green Deal Industrial Plan. This approach will help keep investments in Europe and promote the development of the European ecosystem and the value chain industry, allowing the European

Union's Member States to set subsidies. Other territories and countries are also considering supportive packages for hydrogen production.

3.7.3.2. Clean hydrogen certification schemes

Several methodologies and standards have evolved for certification (and for facilitating comparison) of low carbon hydrogen produced using different methods (e.g. CertifHY [96]). These standards typically define a reference value for the life cycle carbon emissions intensity (or carbon footprint) per unit of hydrogen produced, which enables the hydrogen produced at a particular facility to be qualified or certified as low carbon [97]. Similar certificates are already available for renewable energy in many jurisdictions. Hydrogen certifications may also be expected to be globally accepted and could even become necessary for projects to be eligible for receiving funds from green finance schemes, government incentives, subsidies and concessions towards project development, and even for cross-border trade (e.g. under a carbon border adjustment mechanism proposed in the European Union [98]). The success of such a scheme can be ensured when a harmonized life cycle carbon footprint assessment methodology is adopted by all hydrogen project developers and when it is independently verifiable by certifying bodies.

Owing to the very low life cycle carbon footprint of nuclear energy (which according to a detailed assessment by the United Nations Economic Commission for Europe (UNECE) is even lower than those of the main renewable generation technologies [99]), nuclear assisted hydrogen production would have a unique competitive advantage that would facilitate its entry and participation in the low carbon hydrogen markets.

Table 10 illustrates a simplified carbon emissions intensity calculation of hydrogen derived from water electrolysis from an identical electrolyser plant but using different electricity sources in each case (based on representative data from UNECE [99]). It clearly indicates that nuclear hydrogen can meet existing low carbon emissions standards (which use carbon emissions intensity figures of 2–4 kg of CO_2 per kg of H_2 to certify low carbon hydrogen), thus justifying its inclusion in decarbonization and energy transition plans based on hydrogen. CO_2-e (or carbon dioxide equivalent) is a unit of measurement used to compare the global warming potential of different greenhouse gases by expressing them as the equivalent amount of carbon dioxide.

TABLE 10. TYPICAL LIFE CYCLE CARBON EMISSIONS INTENSITY OF HYDROGEN OBTAINED FROM WATER ELECTROLYSIS USING DIFFERENT ELECTRICITY SOURCES

Electricity generation source	Life cycle CO_2 emissions intensity (g CO_2-e/kWh(e))	Expected CO_2 emissions intensity of electrolytically produced hydrogen (kg CO_2-e/kg H_2)
Coal fired generators (without CCS)	1150	57.50
Natural gas fired generators (without CCS)	660	33.00
Coal fired generators (with CCS)	115	5.75
Natural gas fired generators (with CCS)	66	3.30
World average grid electricity	550	27.50
Hydroelectricity	11	0.55
Solar PV	35	1.75
Wind	13	0.65
Nuclear	5	0.25

Note: Electrical energy consumption in water electrolysis = 50 kWh per kg of H_2 produced; efficiency of CCS = 90% removal of produced CO_2 from the combustion flue gases.

4. TECHNOLOGIES FOR NUCLEAR HYDROGEN PRODUCTION

This section covers various technologies for producing nuclear hydrogen, along with feedwater requirements and regional considerations of water supply. It summarizes and compares the advantages and challenges of the current technologies. Some relevant technical information for the various hydrogen production pathways as well as the ranges of the TRL are included. The section also discusses opportunities for integrating these technologies with nuclear power and the resulting advantages of doing so, where this integration will lead to advantageous hydrogen production. This is not meant to be an exhaustive review of the hydrogen production technologies available and under development, since these types of reviews are provided by many technical reports, such as Ref. [100]. Instead, the intent is to give a brief overview of the technologies for background and context to align with the main purposes of this publication.

4.1. ELECTROLYSIS

Electrolysis, the splitting of water into hydrogen and oxygen, has been performed commercially for a long time, with alkaline electrolysis being the most established technology. The industrial synthesis of hydrogen and oxygen through electrolysis dates back to the end of the nineteenth century.

The mass production of hydrogen through electrolysis has not yet been adopted owing to the large energy requirements and associated costs. As discussed in Sections 2 and 3, the drivers relating to national policies and global goals for decarbonizing the energy economy are catalysing the development of more efficient and environmentally friendly technologies for the production, storage and distribution of hydrogen. As the technology and manufacturing capacity scale up, costs have begun to and will continue to decrease, especially for HTE technologies. Demineralized water and energy are the basic feedstock requirements for electrolysis. Water electrolysis is expected to develop rapidly in the next decade, and by 2030 more than 65% of hydrogen is expected to be produced using electrolysers [101]. Demonstration projects for hydrogen production by electrolysis at existing nuclear power plants are currently on-going. A nuclear power plant's performance, reliability and sustained economic viability have to be addressed when considering hydrogen production using nuclear energy, especially in the context of operational plants, as these could necessitate plant modifications, procedure revisions, new or revised maintenance practices and additional training.

4.1.1. Low temperature electrolysis

4.1.1.1. Description of the technology

LTE encompasses hydrogen production methods at process temperatures lower than 100°C and operates using inputs of electricity and water to produce high purity hydrogen and oxygen. The two main LTE processes are alkaline electrolysis and PEM electrolysis [102]:

- Alkaline electrolysis: This is a historical form of electrolysis where the electrolyser operates in a liquid alkaline electrolyte solution of sodium or potassium hydroxide. These systems are commercially available and have been proven in large scale applications. An advantage of this technology is that it avoids the use of precious metals within the electrodes.
- PEM electrolysis: This is a more recently developed form of electrolysis where a solid polymer electrolyte membrane is used to conduct protons, separate product gases and serve as an electrical insulator to the electrodes.

Although the costs are an important driver and can be an advantage in adopting one technology over another, these costs vary widely and are rapidly changing owing to R&D and commercial developments [102]. Table 11 (based on data from Ref. [102]) summarizes for comparison some of the relevant technical characteristics, advantages and challenges of alkaline electrolysis and PEM electrolysis.

4.1.1.2. Integration with nuclear power

Integration of alkaline electrolysis and PEM electrolysis at a nuclear power facility needs additional infrastructure to supply electricity and water to the electrolysis units. Clean water accessibility at high volumes will be needed to support electrolysis at commercial/industrial scales. Assuming that the electrical connection is made directly with the nuclear unit, additional rectifiers will be needed for converting alternating current to direct current for feeding to the electrolyser. For example, in the USA a new switchyard is expected to be added to avoid associated transmission and regulatory costs before electricity is placed on the grid (referred to as 'behind the meter') [103]. Whether the nuclear power plant is in a regulated or deregulated electricity market will affect the negotiations surrounding behind the meter operations.

As LTE requires only an electrical connection to nuclear power and not a thermal connection, the electrolysis units can be placed further from the nuclear unit, assuming that the electricity used to produce the clean hydrogen can be

TABLE 11. TECHNICAL CHARACTERISTICS, ADVANTAGES AND CHALLENGES OF ALKALINE ELECTROLYSIS AND PEM ELECTROLYSIS

	Alkaline	PEM
Cell temperature (°C)	65–100	70–90
Cell pressure (bar)	25–30	30–80
Charge carrier	OH^-	H^+
Specific system energy consumption (kWh/m³)	4.5–7.5	5.8–7.5
Cathode material	Ni, Ni–Mo alloys	Pt, Pt–Pd
Anode material	Ni, Ni–Co alloys	RuO_2, IrO_2
Voltage efficiency (%)	50.0–70.8	48.5–65.5
Advantages	— The most mature of the existing hydrogen production technologies — Offers lower prices currently and longer lifetimes	— Enhanced dispatchability and cycling time
Challenges	— Electrode corrosion is considered the foremost challenge	— The precious metals used as catalysts add cost, as well as material availability and sourcing concerns for manufacturing at large global scale — Difficulty in scaling up for large scale (MW) applications

attributed to nuclear power and would therefore be eligible to receive subsidies for low carbon production. Additionally, the electrolysis unit will have 'balance of plant' (this refers to all the supporting components and systems necessary for the operation of the main electrolysis unit, excluding the electrolyser stack itself) for converting the voltage and current (alternating to direct current) to the appropriate types for connection to the electrolysers.

The impact of the integration on nuclear power plant operation is minimal, as LTE technologies appear as a standard electrical demand, therefore normal operating procedures apply. Depending on the scale of the electrolysis unit, the hydrogen facility may have a separate control room for operating the system, or it might be just a panel within the electrolysis skid for smaller setups. Interfaces with the nuclear control room will provide alarms and indications and allow nuclear plant operators to divert the power supply from the hydrogen plant for any scenario.

4.1.2. High temperature electrolysis

4.1.2.1. Description of the technology

High temperature steam electrolysis (HTSE), which utilizes solid oxide electrolysis cells (SOECs), operates at temperatures higher than 650°C and uses electricity and steam as a feedstock. The electrical inputs for the SOECs are in two forms, one for obtaining the voltage and current density requirements and the other for resistive heating to achieve the high temperature requirements. HTSE is generally acknowledged to provide an improvement in stack energy conversion efficiency (as measured in kWh input needed per kg of hydrogen produced) compared with LTE technologies because of the higher operating temperatures of the system. This improvement stems from the use of thermal energy (steam) to decrease the amount of electricity needed to achieve the thermal requirements of the SOECs, thereby reducing electricity requirements by 35% compared with LTE [104].

The integration of HTSE with nuclear power offers the ability to use thermal energy from the nuclear plant to provide the needed thermal energy to the HTSE plant, thereby enabling hydrogen production at a higher efficiency compared with PEM and alkaline technologies. The capital costs of HTSE stacks are expected to be lower than for PEM electrolysis — owing to the reduction in precious metal catalysts that are used in PEM electrolysis — and comparable with alkaline electrolysis. However, the costs need to be weighed against the lifetime of the stacks, which is currently significantly shorter for SOECs than for PEM electrolysis. In addition, the integration of hydrogen production using HTSE with nuclear power enables large scale centralized production of hydrogen because of the large amounts of concentrated power available from nuclear power and the higher efficiency of hydrogen production of HTSE. This integration can enable large scale integration of industry in the vicinity of a nuclear power plant in a future industrial energy park [104].

Rapid gains in recent years in the technological development of SOECs — the technology used to implement HTSE — have taken the technology from the laboratory scale to the precommercial stage. Currently, the HTSE/SOEC technology is being tested at scales of a few hundred kW, and lower than 0.5 MW, but it is expected that it will soon be commercialized in the range of >1 MW.

4.1.2.2. Integration with a nuclear power reactor

HTSE has unique characteristics that make it ideally suited for integration with nuclear power. Thermal energy in the form of steam can be harvested from the nuclear reactor turbine cycle. In the case of a pressurized water reactor, the steam is non-radioactive. The steam can be extracted from the turbine cycle and passed through a heat exchanger outside the turbine building to vaporize the demineralized water for the HTSE plant. Once the steam from the turbine cycle has passed through the heat exchanger and is condensed, the condensate is returned to the condenser in the turbine cycle.

While the temperature of the available thermal energy for HTSE can vary, ideally the temperature would be greater than the boiling point of water. Though operating temperatures for HTSE are higher than temperatures that light water reactors can provide, these reactors have heat available in the ranges of 150–300°C. The most valuable use to the HTSE plant of the thermal energy from the nuclear power plant is supplying the heat of vaporization of water, which requires a large amount of energy. After the vaporization of water, the sensible heat required to raise the temperature of the steam from 150–300°C to the operating temperature of 700–850°C can be generated by heat recuperators and electric resistance topping heaters. The recuperators are heat exchangers that recycle the heat from the outlet of the HTSE stacks and transfer it to the feed stream. After vaporization and recuperation, the feed stream temperature is finally raised to the operating temperature of the HTSE plant by electric resistance topping heaters [105].

In the future, the light water reactor SMR designs currently under development and reactors with higher output temperatures, such as high temperature gas cooled reactors (HTGRs), molten salt reactors and liquid metal cooled reactors, could also be used for integration with HTSE hydrogen production. HTGRs could provide a small efficiency gain over existing light water reactors, though likely not a significant one, as they would only displace the need for electric resistance topping heaters. The greatest energy consumption is in vaporizing the feedwater. New reactors could have the advantage of being purpose designed with thermal energy extraction in mind for use in industrial processes such as HTSE hydrogen production.

Analyses of the integration of HTSE hydrogen production with nuclear power are commonly performed with the assumption that the integration would be with a pressurized water reactor. This is because pressurized water reactors have a turbine cycle steam loop separate from the reactor coolant system, which, under normal operating conditions, keeps fission products and activated species separated from the steam extraction. The assumption is that thermal energy for HTSE would be taken from the turbine steam loop.

Boiling water reactors present a unique challenge for HTSE integration, as the turbine steam goes through the nuclear reactor core and, as a result, is minimally radioactive under normal operating conditions. The integration of HTSE with boiling water reactors would still be possible, but it would entail additional complexities or costs, for example for an additional steam loop and heat exchangers to convey the non-radioactive steam to the HTSE hydrogen plant.

4.1.3. Water requirements for electrolysis based hydrogen production

4.1.3.1. Water supply considerations for hydrogen production

Nuclear driven hydrogen production technologies are based on water decomposition (using electricity and/or heat). From the overall stoichiometry of water splitting ($H_2O \leftrightarrow H_2 + 0.5O_2$), it can be calculated that the theoretical minimum water consumption for hydrogen production is about 8.92 kg per kg of H_2 produced [106]. Considering cooling needs for the produced hydrogen, it is estimated that the average total water consumption intensity for hydrogen production by alkaline electrolysis is 22.3 kg of water per kg of H_2 and for PEM electrolysis, 17.5 kg of water per kg of H_2 [107].

For alkaline electrolysis and PEM electrolysis, high purity water (i.e. de-ionized or demineralized grade of water) has to be supplied to the electrolyser at a rate matching the water consumption during electrolysis. Most nuclear plants have on-site demineralized water production capabilities to meet the needs of the plant. It may be possible to utilize the plant's demineralized water production capabilities for hydrogen production, depending on the scale of hydrogen production planned at the site. In case of future expansion of the hydrogen production capacity, demineralized water production facilities would have to be augmented accordingly to meet the increased demand.

For newly built coastal nuclear power plants without access to a freshwater supply, it may be necessary to desalinate seawater to meet the make-up requirement for power production and direct consumption in hydrogen production processes. The make-up requirement refers to the additional water needed to replace losses due to evaporation, leaks, or other consumption during the power production and hydrogen production processes. Thermal desalination processes such as multistage flash or multieffect distillation, with or without vapour compression, may be supported by nuclear steam extracted from the secondary cycle to produce the desired quantities of high purity water. Nuclear generated electricity may also be used to operate membrane based desalination processes (e.g. reverse osmosis) followed by adsorption based techniques using anion and cation exchange resins to produce high quality feedwater for the power plant as well as for the hydrogen production facility.

Research and development (R&D) activities are being pursued in developing direct sea water electrolysers to produce hydrogen. However, technical challenges still remain in the development of the electrolysers (including management of the chlorine released during the process and significantly lower lifetime of the cells in the chloride rich sea water environment), for which further exploratory work (e.g. for preventing chloride crossover) is needed [108].

4.1.3.2. Water for other applications

In addition to the water required for the water splitting reaction, water is also needed for utility functions at the hydrogen plant (i.e. service water), such as for cooling the circulating electrolyte as in the case of alkaline electrolysis. The cooling water may reject heat to the ambient air if the final heat sink is a cooling tower, for which additional water losses in the form of evaporative losses, drift losses and blowdown losses are expected in normal operation. As many nuclear power plants already employ cooling towers on the site for closed loop cooling, the service water for the hydrogen plant may be able to be extracted from the same water circuit.

4.1.4. Conventional steam methane reforming

4.1.4.1. Description of the technology

The steam methane reforming pathway, which is linked to considerable CO_2 emissions (of 7.5–12 t of CO_2 per t of H_2 produced), dominates the global hydrogen generation market [109]. The methane steam reforming process is the catalytic decomposition of light hydrocarbons (e.g. methane, natural gas, naphtha) to react with superheated steam, resulting in a hydrogen rich gas mixture. The reforming reactions are endothermic, running at high temperatures (>500°C). Steam methane reforming typically takes place at 850°C, and at pressures of 2.5–5 MPa, through the endothermic reactions shown in Eqs (1) and (2) [100]:

$$CH_4 + H_2O \leftrightarrow 3\,H_2 + CO\ + 206\ \text{kJ/mol} \quad (1)$$

$$CH_4 + 2\,H_2O \leftrightarrow 4\,H_2 + CO_2\ + 165\ \text{kJ/mol} \quad (2)$$

The carbon monoxide in the first reaction (Eq. (1)) is further catalytically converted in a slightly exothermic water–gas shift reaction with steam to produce additional hydrogen according to Eq. (3):

$$CO + H_2O \leftrightarrow H_2 + CO_2\ - 41\ \text{kJ/mol} \quad (3)$$

4.1.4.2. Integration with a nuclear power reactor

The process requires heat, which is conventionally met by combustion of additional methane. The top temperature range of heating the process is 800–850°C, which can be met by HTGR designs currently under development, saving about 35% of methane feed and a similar amount of CO_2 emission compared to the conventional combustion route. If the CO_2 from the steam methane reforming process is captured, this can be considered a low carbon source of nuclear hydrogen production.

Sponsored by the Ministry of Economy, Trade and Industry and the Ministry of Education, Culture, Sports, Science and Technology, the Japan Atomic Energy Agency (JAEA) is conducting an eight-year national project (awarded the equivalent of US $200 million) to construct and connect a steam methane reformer to the existing 30 MWt High Temperature Engineering Test Reactor (HTTR). The steam methane reformer has a capacity of producing 800 (normal) m^3/h hydrogen, while the HTTR provides the high temperature heat of 750–800°C needed for the process. The aim is to demonstrate the safety design, licensing and long term stable operation of nuclear hydrogen production with the HTGR technology.

4.1.5. Thermochemical cycles

4.1.5.1. Description of the technology

Direct thermal decomposition of water requires the heat of several thousand degrees Celsius, which poses significant challenges in terms of construction material. However, by incorporating chemical reactions in the decomposition process, through the so-called thermochemical cycles, the temperature threshold of the heat required to split water molecules can be significantly lowered.

The iodine–sulphur (I–S) cycle is among the most developed thermochemical processes in the world and Member States such as China,

France, Japan, Republic of Korea and the USA have been investigating it for a long time. Through the three cyclic chemical reactions shown in Eqs (4)–(6), the H_2O is split into H_2 and O_2 products. The iodine and sulphur are reused in the closed cycle process, based on the following chemical sequence [100]:

$$I_2 + SO_2 + 2H_2O \rightarrow H_2SO_4 + 2HI \tag{4}$$

$$H_2SO_4 \rightarrow H_2O + SO_2 + 0.5O_2 \tag{5}$$

$$2HI \rightarrow H_2 + I_2 \tag{6}$$

The first reaction (Eq. (4)) is exothermic, occurring at about 120°C. The other reactions are endothermic, occurring at 800–900°C in the case of the second reaction (Eq. (5)), and at 450°C for the third reaction (Eq. (6)). The heat demand can be met by the HTGR.

At present, the JAEA operates a test facility closed I–S cycle loop, with the capacity to produce ~100 L of H_2/h in normal conditions. The loop has achieved up to 150 h of continuous operation tests. The JAEA is developing additional fluid flow monitor and control technologies with the goal of significantly extending the stable operation period of the loop [110].

4.1.5.2. Integration with a nuclear power reactor

The pre-licensing basic design of an HTGR cogeneration test plant to be coupled with the JAEA's 30 MWt, 950°C HTTR has been completed. More information is given in Section 4.4.3. The test plant simulates the system operation of a JAEA commercial plant design, namely the Gas Turbine High Temperature Reactor for Cogeneration (GTHTR300C). More specifically, the test plant aims to demonstrate that the GTHTR300C system can be licensed for electricity and hydrogen cogeneration, and to confirm the operation and safety performance of such cogeneration systems [111]. The test plant is expected to be the first of a kind nuclear system operating on two of the advanced energy conversion systems attractive for the HTGR technology — a closed cycle helium gas turbine for power generation and a thermochemical I–S process for hydrogen production at a rate of around 30 (normal) m^3/h.

4.1.6. The copper–chlorine hybrid cycle

4.1.6.1. Description of the technology

The copper–chlorine cycle, or the Cu–Cl cycle, is a hybrid cycle consisting of several thermal and one electrochemical reaction. The Cu–Cl cycle is a promising thermochemical cycle for large scale hydrogen production at low temperatures (<550°C). It has several variations. A thermodynamic comparison of the three, four and five step cycles has shown that the four step cycle has the highest energy (41.9%) and exergy (75.7%) efficiencies [112]. Canadian Nuclear Laboratories (CNL) have been developing the four step Cu–Cl cycle in collaboration with Ontario Tech University (formerly the University of Ontario Institute of Technology) and Argonne National Laboratories in the USA [113, 114]. CNL's four step Cu–Cl cycle was trademarked as HCuTEC and is shown in Fig. 20. Hydrogen is produced at the cathode while cuprous chloride (CuCl) is oxidized to cupric chloride ($CuCl_2$) at the anode during electrolysis. The anode stream from the electrolysis step is transferred to the separation step where solid CuCl and $CuCl_2$ are produced, and hydrochloric acid (HCl) is removed. The solid copper species are then transferred to the hydrolysis step where $CuCl_2$ reacts with steam (H_2O) to form copper oxychloride (Cu_2OCl_2) and HCl. The Cu_2OCl_2 product of hydrolysis goes through the thermolysis step to produce CuCl and O_2. CuCl from the thermolysis step and HCl from separation and hydrolysis are then

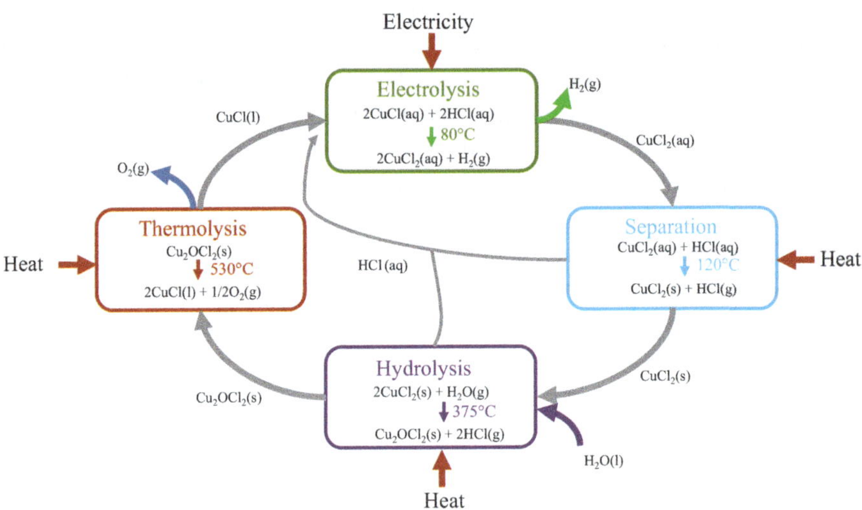

FIG. 20. Schematic of the HCuTEC hydrogen production process.

fed back into the electrolysis step to close the cycle. In recent years, CNL has made significant efforts to advance the electrolysis step. An integrated HCuTEC process was demonstrated in laboratories at a capacity of about 50 L of H_2/h in normal conditions [115]. Further scaling up will involve industrial partners.

Ontario Tech University has also built a laboratory scale Cu–Cl facility. Its development of the Cu–Cl cycle is on-going, with support from CNL and the Canadian Government. The Cu–Cl cycle seems to have attracted interest from other countries as well, for example Bhabha Atomic Research Centre in India [116] and K.N. Toosi University of Technology in the Islamic Republic of Iran [117].

4.1.6.2. Integration with a nuclear power reactor

A nuclear reactor can provide both the heat and electricity requirements for the Cu–Cl cycle. It has been reported in Ref. [118] that the steam methane reforming method has about 12 times more global warming potential than the nuclear based Cu–Cl cycle for hydrogen production. CNL selected the Cu–Cl cycle for hydrogen production owing to its unique advantages; most notably the lower process temperature (<550°C) compared with many other thermochemical cycles; its ability to couple with several types of advanced reactors and SMRs being explored in Canada and around the world; and its potential high overall cycle efficiency. A novel concept was proposed by Ontario Tech University to integrate the Cu–Cl cycle with a supercritical water cooled nuclear reactor for both hydrogen and electricity production [119]. The supercritical fluid (~620°C) directly supplies heat for the $CuCl_2$ hydrolysis and Cu_2OCl_2 decomposition. The energy and the exergy efficiencies of the proposed system are determined to be 31.6% and 56.2%, respectively [119]. A life cycle assessment of nuclear hydrogen production methods has found that the carbon emission (measured as kg of CO_2 equivalent per kg of hydrogen) of the Cu–Cl cycle is more than double that of the SOEC electrolysis [120].

4.1.7. Methane pyrolysis

4.1.7.1. Description of the technology

Methane pyrolysis is the thermal cracking of natural gas directly into hydrogen and solid carbon. For temperatures in the range of 600–900°C, a catalyst is required to speed up the process [121]. The process is hampered by catalyst deactivation due to the solid carbon produced. Above 900°C a catalyst either is not required or the solid carbon product itself can act as a surface catalyst [49]. Three main types of methane pyrolysis have been developed over the past century that all avoid the direct production of CO_2 [122]:

- Catalysed reactions at lower temperature ranges, including the use of a liquid metal or molten salt media;
- Higher temperature thermal cracking, sometimes through induction heating;
- Thermal and electrochemical methods (i.e. arc or microwave plasma splitting).

While methane pyrolysis requires moderately more natural gas (as a feedstock) per kg of hydrogen produced than steam methane reforming, it has the significant advantage of requiring the lowest amount of energy (lower than steam methane reforming and about eight times lower than water electrolysis). At the same time, it produces no direct CO_2 emissions; it produces only solid carbon, which can be a valuable by-product to be used for various chemical feedstocks [123].

Arc plasma techniques have been commercially employed for decades, but when applied to methane pyrolysis they are not nearly as energy efficient as other pyrolysis methods under development. Other techniques are significantly less developed, but owing to their low theoretical energy requirements, they have the potential to provide cost competitive hydrogen production [124]. Many institutes and companies are putting extensive effort into developing various methods of methane pyrolysis.

4.1.7.2. Integration with a nuclear power reactor

For higher temperature pyrolysis methods that go above 900°C, the predominant energy source is electricity, which could be provided by current or advanced nuclear reactors at a cost competitive and scalable level. Of even greater interest, however, is the fact that many techniques require energy at a low enough temperature as to be provided directly as thermal energy from high temperature nuclear systems. Without the need for conversion to electricity, a nuclear plant will produce 2–3 times more thermal energy than net electrical. Thus, there is great potential for advanced high temperature reactors to produce large quantities of low cost hydrogen. Methane pyrolysis is an attractive process, as it is expected to have the lowest possible cost per kilogram of hydrogen produced.

4.1.8. Comparison of technologies for nuclear hydrogen production

4.1.8.1. Assigning technology readiness levels to nuclear hydrogen production processes

The basis of assigning a technology readiness level (TRL) and a commercial readiness index (CRI) value to a particular hydrogen production technology suitable for coupling with a nuclear reactor (current generation/advanced/SMRs) may be understood from Fig. 21 (adapted from Ref. [125]), which provides a qualitative interpretation corresponding to each numerical value on the scale designating the level. These values indicate the maturity level of a process or technology and provide an indication of its suitability for a full scale commercial application catering to an industry sector's needs. CRI values above 2 normally become applicable only for technologies that have attained TRLs of 7 and above, since, at this level, individual components may be considered sufficiently well developed to be integrated into systems suitable for commercial use.

For hydrogen technologies, factors such as the following can be taken as an indication of a high TRL (i.e. 8–9) and a high CRI (i.e. 5–6) for that particular technology:

- Availability of multiple vendors (in multiple regions) for the same technology (e.g. hydrogen production system, balance of plant components, power systems and electronic components, materials for construction, safety systems, sensors, detectors);
- Availability of reliable technical and commercial data and prior operational experience for the technology option;
- Availability of a reliable supply chain of process or plant components for regular maintenance, replacements and facility upgrades;
- Examples of announced projects with final investment decisions made for specific technologies for commercial scale hydrogen production;
- Well defined users/offtake industries for the produced hydrogen with well developed infrastructure to transport and store the produced hydrogen;
- Examples of further innovation and technology upgrades being taken up by industrial organizations.

These technologies (with high TRL and high CRI) can therefore be available for coupling with the current generation of nuclear reactors for nuclear hydrogen production. Low carbon hydrogen production processes currently include low temperature electrolysis technologies such as alkaline electrolysis and PEM electrolysis. The HTSE processes and anion exchange membrane electrolysis are gradually gaining market share.

Hydrogen production technologies for which lower TRL values have to be assigned have the following characteristics:

— There is current interest from academia and research organizations.
— Technology development studies at the laboratory scale are in progress.
— Substantial technical and commercial uncertainties exist, which need to be resolved by further demonstration projects.
— There is a need for a scale up of systems and components to meet industrial user demands.

Commercial readiness is also correspondingly at a low level for these options. Thermochemical cycles for hydrogen production may be considered to belong to this group of process technologies.

Based on this rationale, Table 12 shows possible assignments of TRLs and CRIs to various low carbon hydrogen production technologies currently under consideration.

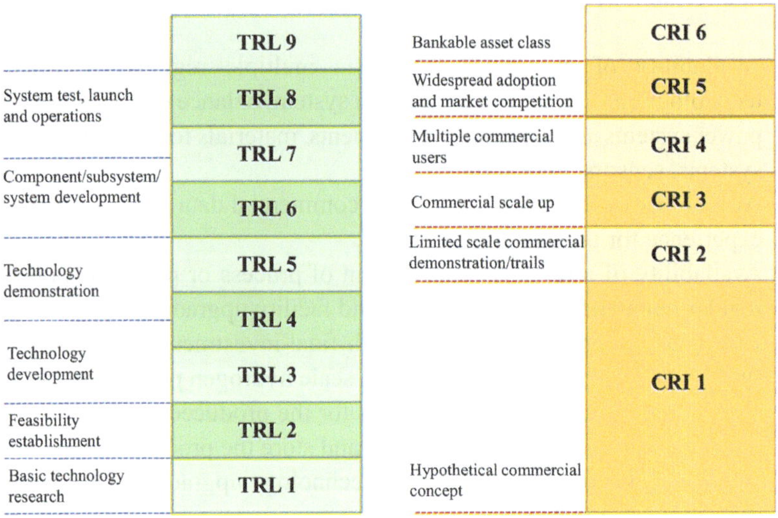

TRL 9
System test, launch and operations — TRL 8
TRL 7
Component/subsystem/ system development — TRL 6
Technology demonstration — TRL 5
TRL 4
Technology development — TRL 3
Feasibility establishment — TRL 2
Basic technology research — TRL 1

Bankable asset class — CRI 6
Widespread adoption and market competition — CRI 5
Multiple commercial users — CRI 4
Commercial scale up — CRI 3
Limited scale commercial demonstration/trails — CRI 2
CRI 1
Hypothetical commercial concept

FIG. 21. General explanation of technology readiness levels and commercial readiness index values for technologies (adapted from Ref. [125]).

TABLE 12. ASSIGNING TECHNOLOGY READINESS LEVELS AND COMMERCIAL READINESS INDEX VALUES TO NUCLEAR HYDROGEN PRODUCTION TECHNOLOGIES

Production technology	TRL	CRI
Alkaline water electrolysis	9	6
PEM electrolysis	9	6
SOEC electrolysis	6–7	2–3
Anion exchange membrane electrolysis	6	3–4
Intermediate temperature thermochemical cycles	4–5	1–2
High temperature thermochemical cycles	4–5	1–2
Methane pyrolysis	4	1–2

4.2. NUCLEAR HYDROGEN IN HYBRID ENERGY SYSTEMS

The production of hydrogen with nuclear power offers a number of opportunities for nuclear plant owners and operators as they look towards the future and their need to participate in a grid with high renewable penetration. Nuclear plants, depending on their locations, have either operated as a baseload and provided 100% of reactor power to the grid continuously, or operated flexibly to match grid requirements. Recent years have seen a tremendous growth of renewables, which has tended to push an increasing number of nuclear units away from baseload operation into flexible operations. Given the cost structure of commercial nuclear power, which is dominated by O&M costs and CAPEX, reducing power has little or no impact on fuel or the cost of operating the plant.

Another factor impacting historical nuclear plant baseload operations is the market fluctuations in natural gas pricing. In countries where natural gas prices are low due to large supply or other factors, natural gas becomes a more attractive option for electricity generation, creating a challenging environment for nuclear power plants.

However, hydrogen production offers an opportunity for nuclear units to maintain reactor power at 100% and to divert power between hydrogen production and supplying the grid. Hydrogen can thereby act as a long term energy storage mechanism (albeit with a lower round trip efficiency due to additional conversion and reconversion steps) or as a feedstock for other industrial applications such as ammonia production, synthetic fuel production, substitute natural gas, petroleum refining and other chemical processes. This allows the grid to accept more renewables while having the nuclear plant operating in optimal condition at 100% power.

Higher temperature reactors will offer new opportunities and methods for hydrogen production, as on-going efforts in several countries have shown. For example, in the USA the Department of Energy investigated the design and licensing of HTGRs for the production of hydrogen through the Next Generation Nuclear Plant project [126].

Aside from hydrogen, nuclear reactors are poised to play an increasing role in large scale decarbonization of process and manufacturing industries.

4.3. TECHNICAL ASPECTS OF HYDROGEN TRANSPORATION AND STORAGE

The hydrogen produced at the nuclear reactor site would need to be transported to the location of the end user. Depending on the location of the user and the quantity of hydrogen to be dispatched, various transportation options have to be considered [127, 128]. Depending on the application, varying sizes of intermediate gas storage systems may also be required. Tables 13 and 14 provide an overview of these technologies and some considerations of relevance to nuclear hydrogen production projects.

TABLE 13. OVERVIEW OF HYDROGEN TRANSPORTATION TECHNOLOGIES

Technology	Typical technical specifications	Relevance to nuclear hydrogen production projects
Gas pipeline	High alloy metal with or without polymeric lining, diameter up to 1.22 m, gas pressure up to 70 bar, gas linear velocities up to 30 m/s, pipeline lengths of few hundred to thousand km. The amount of hydrogen allowed in existent gas pipelines is under investigation	Can be utilized for continuous transport of hydrogen produced at the reactor site to the industrial/commercial user located beyond the exclusion zone (typical distances >20 km)

TABLE 13. OVERVIEW OF HYDROGEN TRANSPORTATION TECHNOLOGIES (cont.)

Technology	Typical technical specifications	Relevance to nuclear hydrogen production projects
Tanker, tube trailer	On board gas pressure ~700–800 bar, suitable for transport over distances of few hundred km from point of hydrogen production; may also be suitable for transport of liquid phase derivatives of hydrogen like ammonia, methanol or liquid organic hydrogen carriers	Can be used for localized distribution of nuclear produced hydrogen to users who have provisions for on-site hydrogen storage and/or for users who require hydrogen intermittently
Barge, ship	Suitable for transport of bulk quantities of hydrogen (gaseous or liquid) or hydrogen derivatives such as ammonia and methanol over a distance of a few thousand km (i.e. trans-continental/oceanic distances)	Can be used for export/trade of nuclear produced hydrogen/ hydrogen derivatives

TABLE 14. OVERVIEW OF HYDROGEN STORAGE TECHNOLOGIES

Technology	Typical technical specifications	Relevance to nuclear hydrogen production projects
Gas storage in natural geological formations	Depending on nature of geological formation (e.g. salt caverns), gas storage pressure may be up to about 30 bar; capacities may be tens to hundreds of thousands of cubic metres of hydrogen; storage duration from days to months	Can act as primary or buffer storage for nuclear produced hydrogen at reactor/end user site, depending on availability of suitable locations

TABLE 14. OVERVIEW OF HYDROGEN STORAGE TECHNOLOGIES (cont.)

Technology	Typical technical specifications	Relevance to nuclear hydrogen production projects
Gas storage in engineered structures	Underground or overground engineered structures for gas storage, in regions where natural formations are not available; long distance pipelines can also act as engineered gas storage system; storage duration from days to weeks	Can act as primary or buffer storage for nuclear produced hydrogen at reactor/end user site
Liquid hydrogen storage	Cryogenic hydrogen storage in liquefied form at –253.15°C temperature in Horton spheres or bullets; storage duration from days to weeks; typical boil off losses of about 0.5% of stored volume per day	Can be part of bunkering arrangements at ports that will trade/export hydrogen by ship/barge
High pressure gas storage vessels	Suitable for limited quantities of storage; typical pressures of 350–700 bar; storage duration from days to weeks	Can be part of small scale storage system (e.g. storage of hydrogen for NPP turbine cooling applications)

Note: NPP — nuclear power plant.

4.4. NUCLEAR HYDROGEN DEMONSTRATION

Nuclear powered hydrogen production offers the opportunity to produce clean hydrogen with a high capacity factor, reliability and scale in centralized locations where nuclear power plants are usually sited near existing industrial centres or in areas with available land for industrial expansion. This makes nuclear integrated hydrogen production attractive if it can be produced with competitive economics. Barriers to nuclear produced hydrogen (and hydrogen production for general use) include economics, technology and the supply chain issues inherent in any new industry development. Various demonstrations at nuclear power plants around the world are under way to prove the concept and the economics of nuclear powered hydrogen production. The utility demonstration projects, as of the end of 2022, are included in Ref. [4]. Some developments are included as examples below.

4.4.1. Canada

In Canada, Bruce Power is investigating the opportunity to produce hydrogen using nuclear energy. In 2022, Bruce Power signed a memorandum of understanding with Hydrogen Optimized, Greenfield Global and Hensall Co-op to develop a feasibility study to explore opportunities for hydrogen production at the Bruce Power site [129]. Information on the nuclear hydrogen roadmap in Canada is provided in Annex A–1.

4.4.2. European Union

The European Union is funding a feasibility study on nuclear powered hydrogen cogeneration under the NPHyCo project[3]. The project receives funding from the Euratom Research and Training Programme (2021–2025) and is investigating hydrogen production using nuclear electricity to address the European Union's challenge to fully decarbonize its economy by 2050. NPHyCo started in the second half of 2022 and will finish in the first quarter of 2025. The project is focusing on the potential for developing large scale, low carbon, hydrogen production facilities linked to nuclear power plants, including the feasibility of producing hydrogen near an existing nuclear power plant and the added value of such a project.

The project is investigating different scenarios for different levels of integration of the hydrogen production plant and the nuclear power plant and aims to deliver an educated proposal for a suitable level of integration and a map of European nuclear power plants where conditions for building NPHyCo plants are favourable, including potential locations for the implementation of a pilot project [130]. Some relevant input data for the study conducted by NPHyCo are given in Table 15.

In Sweden, the Swedish power company OKG AB signed with Linde Gas AB its first contract in 2022 to supply hydrogen produced at its Oskarshamn nuclear power plant [131]. Vattenfall has also been producing hydrogen using electricity from Ringhals nuclear power plant for its own use since 1997.

4.4.3. Japan

A project to license the HTTR coupled with a hydrogen production plant is on-going in Japan. The project will establish the technology necessary for large scale, low cost, carbon free hydrogen production by the JAEA in collaboration with Mitsubishi Heavy Industries [132].

---

[3] https://www.nphyco.org

TABLE 15. INPUT DATA FOR THE STUDY CONDUCTED BY NPHyCo

Characteristic	Value
Power	30 MW/100 MW
Technology	Low temperature electrolysis
Site	European NPP site locations
Hydrogen production potential	12–40 t/d
Year of finishing feasibility study	2025
Year of building first NPHyCo pilot plants	<2030
Type of reactor	Various existing NPP reactor types in Europe
Hydrogen usage	Various industries and European hydrogen pipeline backbone

Note: NPP — nuclear power plant.

Since 2022, work has begun to establish a safe connection technology, including licensing by the Nuclear Regulation Authority, to couple the HTTR with a hydrogen production facility and use the high temperature heat (>900°C) of the HTTR for hydrogen production.

The project contains the following specific activities:

— HTTR nuclear hydrogen production test: The existing HTGR test reactor — the 950°C, 30 MWt HTTR — is to be connected to a hydrogen production facility that will be based on a natural gas steam reforming method. The test aims to license the coupling of the conventional hydrogen production facility with the nuclear reactor for hydrogen production using the high temperature nuclear heat supplied by the HTTR. This aims to demonstrate the regulatory approval for the major coupling equipment technologies, including the high temperature gas isolation valve, high temperature piping and gas circulator that are to be developed. The test also aims to complete the safety design standards and safety licensing requirement for connecting an HTGR to a high temperature hydrogen production facility, not only steam reforming plants but also other thermally coupled plants including, for example, thermochemical cycles and HTE.

— Conceptual design for commercial sized coupling equipment: Based on the test plant, the scale up and conceptual design for feasibility validation of the major coupling equipment, including high temperature helium piping, high temperature helium isolation valves and helium circulators, will be carried out.
— Feasibility study of carbon free hydrogen production technologies: Several promising clean hydrogen production methods, including steam reforming of methane with CCS, thermochemical I–S process, HTE and methane pyrolysis, are planned to be investigated and the optimal method will be selected for use in future commercial nuclear hydrogen production with the HTGR technology.
— Full scale hydrogen production: The Green Growth Strategy through Achieving Carbon Neutrality in 2050 (released in June 2021) indicates that the HTTR is expected to produce large quantities of carbon free hydrogen after 2030 [133].

4.4.4. Russian Federation

The State Atomic Energy Corporation "Rosatom" is focused on low carbon hydrogen production for a wide range of industrial consumers. Among the production technologies under consideration are LTE coupled with a nuclear power plant or renewable source of energy, and steam methane reforming with CCUS. Targeting large scale hydrogen production, ongoing R&D includes development of the hydrogen production supply chain (e.g. production technology, storage and transportation, analysis of potential areas of application). The first testing complex is reported to be at the Kola nuclear power plant site with nuclear hydrogen production (using LTE) mostly for local Russian projects with a focus on decarbonization of industrial facilities [134]. The technical characteristics of the project are presented in Table 16.

TABLE 16. TECHNICAL CHARACTERISTICS OF THE HYDROGEN PROJECT AT THE KOLA NUCLEAR POWER PLANT

Characteristic	Value
Power	1 MW, with possible increase to 10 MW
Technology	Low temperature electrolysis
Site	Kola nuclear power plant
Hydrogen production potential	Up to 25 000 t/a
Year of commissioning	Expected 2025
Type of reactor	WWER-440
Main hydrogen usage	Production of synthetic fuels

TABLE 17. TECHNICAL CHARACTERISTICS OF THE HYDROGEN PROJECT IN THE UST-YANSKY DISTRICT

Characteristic	Value
Power	200 MW
Technology	Electrolysis
Site	SMR nuclear power plant in the Ust-Yansky District
Hydrogen production potential	—[a]
Year of commissioning	2028
Type of reactor	RITM-200
Hydrogen usage	Various industries

[a] —: not available.

Another project at the R&D stage is the construction of the SMR based power plant in the Ust-Yansky District (Republic of Sakha, Yakutia). The SMR construction (based on RITM-200 reactors) in Yakutia is planned to be completed by 2028. The SMR nuclear power plant can be used for low carbon hydrogen production for the various industries. The new SMR will replace coal fired and diesel facilities in the Ust-Yansky District and will provide clean power to the Kyuchus gold deposit development project in the north of the Verkhoyansky District, Yakutia [135]. Technical characteristics of the project are presented in Table 17.

4.4.5. United States of America

In the interest of catalysing the development of hydrogen production technologies and the associated manufacturing base and supply chain, the Department of Energy has awarded funding to various large nuclear utility companies to create nuclear integrated demonstration projects. These projects vary in supplier, technology, use case and market location, and test state of the art electrolysis technologies, paving the way for integration of hydrogen production using nuclear power. Beyond technology demonstration, these projects will also pave the regulatory and safety paths for documenting compliance with industrial safety regulations and best practices.

The nuclear hydrogen demonstration projects include three LTE and one HTE demonstrations. The companies demonstrating LTE are Constellation Energy, Energy Harbor and Arizona Public Service. Xcel Energy will be demonstrating HTSE [136].

These projects have been highlighted through the Department of Energy Hydrogen Program, H2@Scale initiative and the Annual Merit Reviews [137–139]. A short summary of these demonstrations and their status as of September 2023 is provided below for context (see also Ref. [4]):

- Constellation Energy Nuclear Hydrogen LTE Demonstration: Constellation Energy has partnered with Nel Hydrogen to commission an LTE plant of approximately 1 MW(e) at Nine Mile Point Nuclear Station in Oswego, New York. Hydrogen production started in early 2023 [140], aiming to be utilized on-site for pre-existing needs using existing hydrogen storage and supporting infrastructure. Separately, Constellation has partnered with the New York State Energy Research and Development Authority to demonstrate a fuel cell facility.
- Energy Harbor Nuclear Hydrogen LTE Demonstration [141]: Energy Harbor is being funded by the Department of Energy to develop an LTE based nuclear hydrogen demonstration at the Davis–Besse Nuclear Power Station outside of Toledo, Ohio. The Davis–Besse project is proceeding with an engineering design to connect a new hydrogen switchyard to the existing power transmission switchyard to dispatch power to a 2 MW PEM electrolysis unit. The hydrogen switchyard will allow future expansion of hydrogen production of up to 60 MW electrical capacity.
- Xcel Energy Nuclear Hydrogen HTSE Demonstration: Xcel Energy (Northern States Power) and its collaborators are developing an HTSE demonstration at their Prairie Island Nuclear Generating Plant in Red Wing, Minnesota. This demonstration will be unique, in that it will seek to test the use of not only nuclear electricity but also nuclear heat for integration with

HTSE, to take advantage of the efficiency gains in hydrogen production that can be realized using HTSE with a large heat source. The installation will use steam and electricity from the nuclear plant to operate a 240 kW HTSE system capable of producing approximately 125 kg of hydrogen per day, a first of its kind.

- Arizona Public Service LTE Demonstration — Hydrogen for Energy Storage/Arbitrage: Arizona Public Service is considering an LTE hydrogen demonstration that would use hydrogen as an energy storage medium to balance the grid market by energy arbitrage. Negotiations with the Department of Energy on the award and demonstration specifics are on-going.

5. SAFETY AND REGULATORY ASPECTS

The objective of this section is to recall the hydrogen properties related to safety and to provide an overview of safety regulations, codes and standards commonly used in non-nuclear industries to address the hydrogen explosion risk.

5.1. INTRODUCTION

In view of several nuclear accidents, including at the nuclear power plants at Three Mile Island (1979) and more recently at Fukushima Daiichi (2011), hydrogen safety is one of the most relevant topics in reactor safety research and has been addressed in numerous national and international research projects. The significant development of the state of the art regarding hydrogen safety at nuclear power plants has been reflected in relevant publications, such as IAEA-TECDOC-1661, Mitigation of Hydrogen Hazards in Severe Accidents in Nuclear Power Plants [142], and the OECD Nuclear Energy Agency's Status Report on Hydrogen Management and Related Computer Codes [143]. However, safety and regulatory aspects relating to nuclear hydrogen production are a different story, as they involve different sets of regulations and are even built on knowledge from different research and expert communities. Unlike in the case of nuclear accident scenarios, nuclear hydrogen production takes place outside the reactor, outside the containment and, in some cases, far from the nuclear power plant site.

Furthermore, there is a fundamental difference in the safety design philosophies of nuclear power plants and hydrogen production facilities.

For nuclear power plants, the safety regulations and the mitigation measures implemented aim to limit the consequences of an accident, such as combustion loads and possible fission product releases. In contrast, hydrogen production facilities prevent the accumulation of flammable gas mixtures by enabling ventilation and dilution, thus avoiding containment and congestion.

It is crucial for the successful implementation of nuclear hydrogen production to comply with the regulations for both nuclear power and hydrogen production technologies, which have their own specificities. Owing to the specific safety relevant considerations of producing hydrogen, the prevention of hazards connected with hydrogen production, storage and distribution is essential for the widespread acceptance of nuclear hydrogen production. Any failure could damage the public's perception of both nuclear power and hydrogen production, with considerable consequences for the public and for the related industries.

The risks associated with hydrogen production are dependent on the technology used. The most common risks of hydrogen production using nuclear energy can be grouped as follows, based on the hydrogen production technology:

— Hydrogen from fossil fuel processing (e.g. steam methane reforming, coal gasification, partial oxidation of hydrocarbons). Combustible gases (e.g. hydrogen, carbon monoxide, methane, other hydrocarbons) are involved in the process, either as reactants or reaction products. Any leakage or release of these gases into the production unit may pose safety concerns (e.g. asphyxia among workers, structural damage from an explosion).
— Hydrogen from water splitting by electrolysis (low or high temperature). The risk of a hydrogen explosion cannot be excluded. The hydrogen contained in the system can be released through leakage of electrolyte following overpressure, or through failure of the cooling water system.
— Hydrogen from water splitting by thermochemical cycles (e.g. I–S cycle). This technology incorporates chemical reactions into the water decomposition process to produce hydrogen. Hazardous chemicals are involved. Some are commonly used in chemical processing and manufacturing. In this case, the hazards to be considered include loss of containment of highly corrosive substances, loss of containment of toxic gases, loss of containment of flammable gases, and the enhanced flammability of materials in pure oxygen.

Depending on the technology used for hydrogen production, safety requirements of both parts — nuclear power plant and hydrogen production facility — can be considered coupled to a different extent. A crucial parameter here is the distance between the hydrogen production facility and the nuclear power plant. For example, while the production of hydrogen by LTE with electrical power generated by the nuclear power plant allows complete decoupling of both sites (and their facilities), the application of thermochemical cycles involves close interdependencies of both nuclear and hydrogen safety aspects.

5.2. HYDROGEN PROPERTIES AND SAFETY RELEVANT PHENOMENA

Most of the physical properties of hydrogen gas differ significantly from those of other comparable gases. Being a flammable gas, these properties are associated with a specific safety relevance. Owing to its very low density (14 times lighter than air [144]), hydrogen leaks can occur more easily than other gases. In addition, hydrogen has a very wide ignition range (about 7 times wider than for methane [144]), which means that flammable mixtures with air can form more easily. For hydrogen–air mixtures, the ignition energy is lower by a factor of 10 than for mixtures of natural gas and air, for example [145]. The high flame speed (an order of magnitude higher than gasoline) enables an easier transition to detonations and high combustion pressures [146]. However, not all properties of hydrogen contribute to an increased risk. For example, the high diffusivity ensures that hydrogen can be easily dispersed and can therefore be diluted more quickly [146]. Fast combustion combined with the buoyant behaviour can even lead to less severe fire scenarios than for other fuels [147].

Hydrogen accumulation may lead to flammable cloud formation that induces pressure and temperature loads in case of a combustion. Hydrogen has a large flammability domain ranging from a concentration of 4 vol. %, which corresponds to the lower limit of flammability in dry air at 100 kPa and 25°C, and 73 vol. %, which corresponds to the upper limit in the same conditions. The flammability limits depend not only on the gas composition but also on the initial temperature and pressure [148, 149]. The effect of the initial temperature on the hydrogen with a lower flammability limit in air is represented in Fig. 22 [150], based on data from Refs [151, 152] and the effect of the initial temperature on the hydrogen with a lower and upper flammability limit in air is represented in Fig. 23 [150], based on data from Refs [151, 153].

Flammable hydrogen clouds may be ignited in several ways, for example by electrical discharge, gas compression, hot surface and hot jets [100, 154]. In all cases, the energy delivered is rather low, ranging from microjoules to a few joules, and the surface temperature threshold is around 580°C.

After ignition, and depending on gas composition, turbulence level and geometrical configuration, the flame speed may increase from a few metres per second to several hundred metres per second, inducing then dynamic pressure loads that may endanger the surrounding structures as well as facilities in the neighbourhood.

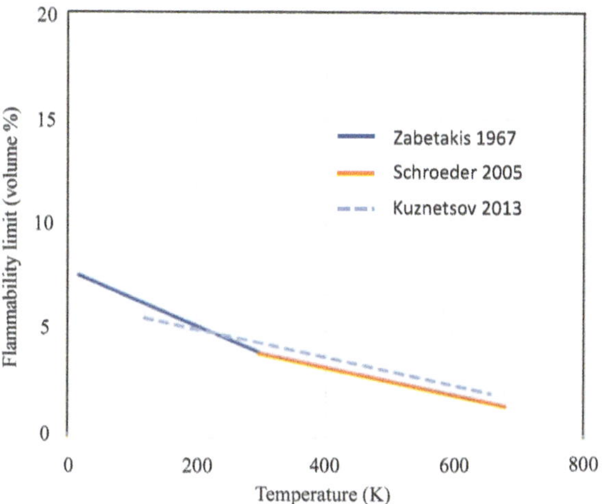

FIG. 22. Effect of initial temperature on hydrogen lower flammability limit in air (adapted from Ref. [150]; see Refs [151, 152] for information contained in the figure).

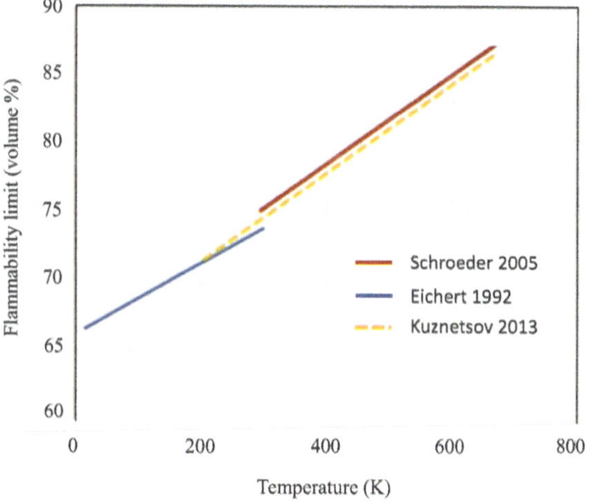

FIG. 23. Effect of initial temperature on hydrogen upper flammability limit in air (adapted from Ref. [150]; see Refs [151, 153] for information contained in the figure).

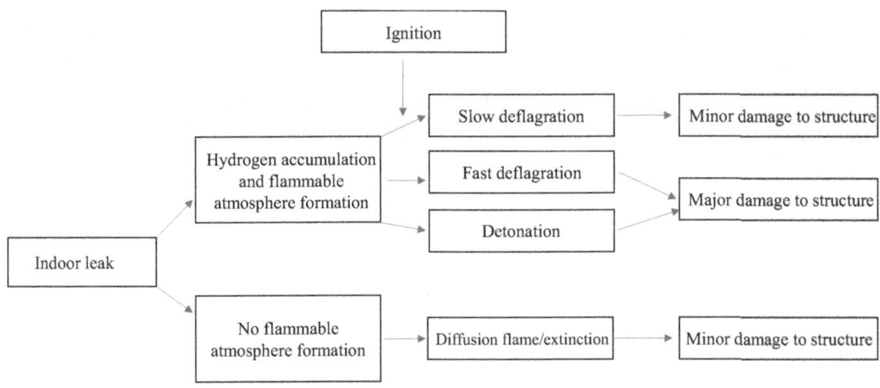

FIG. 24. Generic accident scenarios for leaked hydrogen.

Figure 24 presents a schematic overview of the possible scenarios of a generic hydrogen accident involving leaked hydrogen. These scenarios are relevant for hydrogen production, hydrogen storage and hydrogen transportation.

As for any flammable gases, the primary goal of safety measures relating to hydrogen is the prevention of flammable gas mixtures. Owing to the low ignition energy, the exclusion of ignition sources proves to be very difficult for potential hydrogen–air mixtures. To prevent hydrogen combustion and limit its consequences, regulations, codes and standards (RCSs) are used in the chemical and nuclear industries. Mitigation measures are then implemented accordingly. The most used RCSs are listed in the next section.

5.3. REGULATIONS, CODES AND STANDARDS, AND BEST PRACTICES

In nuclear power plants, hydrogen is used in several technical processes as part of normal operational routines. In the case of a severe accident, the most significant hydrogen risk comes from core oxidation (as was the case in the accident at the Fukushima Daiichi nuclear power plant [142]). Hydrogen released as a consequence of a severe accident with core oxidation creates a challenge for containment as the last barrier against radioactive release gets released into the environment and can only be mitigated by dedicated measures (e.g. passive auto-catalytic recombiners, igniters). In the case of hydrogen used as a technical gas for turbine cooling, the primary associated hazards are combustion, pressure, low temperature, hydrogen embrittlement and exposure. Here, specifically designed directions and guidelines for controlling explosive atmospheres apply.

For example, in Europe, such guidelines are included in two European directives known as ATEX directives (name derived from the French term 'Atmosphères Explosibles') [155]. Directive 99/92/EC [156] (ATEX 153 (formerly ATEX 137) or ATEX Worker Protection Directive) and Directive 2014/34/EU [157] (ATEX 114 or ATEX Equipment Directive) are implemented to ensure safe operation. While ATEX 153 defines minimum requirements for improving the safety and health protection of workers potentially at risk from explosive atmospheres, ATEX 114 provides regulations for equipment and protective systems intended for use in potentially explosive atmospheres. Similarly, in the USA, the National Fire Protection Association standard NFPA 497 [158] provides criteria to determine ignitability hazards in chemical process areas using flammable liquids, gases or vapours, which will assist in the selection of electrical systems and equipment for safe use in Class I hazardous (classified) locations.

Safe hydrogen production considers relevant international, national and regional RCSs. The regulations establish the minimum safety requirements, which are based on well established methodologies; the current state of practice, which also reflects the established standards, refers to best practices, guidelines and industry codes. Regulations are revised as additional new knowledge and insight is gained [159].

In all cases, RCSs aim to provide a framework to prevent accidents and to limit the frequencies of potential hazardous events. The applicability of different RCS frameworks depends on the respective national legislation. Some standards, such as those issued by the International Organization for Standardization (ISO), are internationally accepted rules with a large regional distribution. In Japan, there is no specific regulatory framework for hydrogen. However, hydrogen is subject to existing international regulations relating to dangerous gases in each stage of the supply chain. Owing to the explosivity and combustibility of hydrogen, the High Pressure Gas Safety Act of 1951 introduced regulations regarding the production and storage of hydrogen [160, 161]. According to Ref. [162], China has not yet established specific or unified laws or administrative regulations for hydrogen energy use. The development of hydrogen energy in China is primarily based on national industrial planning policies and local pilot regulations.

The potentially applicable RCSs are given in the next sections.

5.3.1. Hydrogen related regulations, codes and standards

Standards for using hydrogen technologies are made by global standardization bodies (e.g. ISO, American Society of Mechanical Engineers, International Electrotechnical Commission), or regional ones (e.g. European Committee for Standardization, European Committee for Electrotechnical Standardization). For ISO, technical committee 197 on hydrogen technologies has the leading role for the development of standards relating to general handling of hydrogen, and hydrogen production, storage and transportation [163]. RCSs for areas such as pressure vessels, pipelines and gas quality have to be taken into consideration. Some relevant examples are:

- Directives related to explosive atmospheres (e.g. ATEX Directive 99/92/EC [156]);
- Stationary and transportable pressure vessels regulations (e.g. ASME Boiler and Pressure Vessel Code 2021 [164], Pressure Equipment Directive 2014/68/EU (stationary) [165], Transportable Pressure Equipment Directive 2010/35/EU (transportable) [166]);
- Transportation of Dangerous Goods (United Nations Model Regulations, the so-called Orange Book) [167].

5.3.1.1. Regulations, codes and standards for general hydrogen safety

The following RCSs are among the most relevant ones relating to the general use of hydrogen. These RCSs are already implemented in the context of technical use of hydrogen in nuclear power plants.

- NFPA 2 Hydrogen Technologies Code [168]. It provides the fundamental safeguards for the generation, installation, storage, piping, use and handling of hydrogen in compressed gas form or cryogenic liquid form.
- ISO/TR 15916:2015: Basic Considerations for the Safety of Hydrogen Systems [169]. It provides guidelines for the use of hydrogen (gaseous and liquid) as well as its storage in either of these or other forms (hydrides). It also includes the basic safety concerns, hazards and risks based on the properties of hydrogen that are relevant to safety.
- IEC 60079:2022 SER Series, Explosive Atmospheres [170]. This includes classification of hazardous areas, inspection and maintenance.

5.3.1.2. Regulations, codes and standards for hydrogen production

The following RCSs are relevant for hydrogen production sites:

- ISO 16110-1:2007: Hydrogen Generators Using Fuel Processing Technologies, Part 1: Safety [171]. This applies to hydrogen generation systems with a capacity of less than 400 m^3/h at 0°C and 101 325 kPa that convert an input fuel to a hydrogen rich stream of composition and conditions suitable for the type of device using the hydrogen (e.g. a fuel cell power system, a hydrogen compression, storage and delivery system).

It also applies to hydrogen generators using, for example, natural gas and other methane rich gases derived from renewable (biomass) or fossil fuel sources, fuels derived from oil refining.

- ISO 22734:2019: Hydrogen Generators Using Water Electrolysis — Industrial, Commercial, and Residential Applications [172]. This includes the definition of the construction, safety and performance requirements of modular or factory-matched hydrogen gas generation appliances using electrochemical reactions to electrolyse water to produce hydrogen.
- ISO/TS 19883:2017: Safety of Pressure Swing Adsorption Systems for Hydrogen Separation and Purification [173]. This establishes the safety measures and applicable design features that are used in the design, commissioning and operation of pressure swing adsorption systems for hydrogen separation and purification, processing all kinds of impure hydrogen streams as feed.
- The quality of hydrogen produced should be also regulated, with specific considerations if further processing is required, for example, ISO 14687:2025: Hydrogen Fuel Quality — Product Specification [174] indicates the minimum quality characteristics of hydrogen fuel as distributed for use in different applications.

5.3.1.3. Regulations, codes and standards for hydrogen storage and transportation

The following are examples of RCSs for hydrogen storage and transportation. They consider pressurized gaseous hydrogen, cryogenic/liquid hydrogen, and liquid organic hydrogen carriers:

- ASME B31.12-2023: Hydrogen Piping and Pipelines [175]. This standard includes specification of requirements for piping in gaseous and liquid hydrogen service and pipelines in gaseous hydrogen service, covering materials, brazing, welding, heat treating, forming, testing, inspection, examination, operation and maintenance.
- ISO 21029-1:2018: Cryogenic Vessels — Transportable Vacuum Insulated Vessels of not more than 1000 Litres Volume, Part 1: Design, Fabrication, Inspection and Tests [176]. This standard outlines the requirements for the design, fabrication, type testing and initial inspection and testing of transportable vacuum insulated cryogenic pressure vessels with a capacity of up to 1000 litres.
- ISO 21029-2:2015: Cryogenic Vessels — Transportable Vacuum Insulated Vessels of not more than 1000 Litres Volume — Part 2: Operational Requirements [177]. This standard includes operational requirements for

transportable vacuum insulated cryogenic vessels with a volume of not more than 1000 litres designed to operate above atmospheric pressure, including putting into service, filling, withdrawal, transport within the location, storage, maintenance, periodic inspection and emergency procedures.

- ISO 20421-1:2019: Cryogenic Vessels — Large Transportable Vacuum Insulated Vessels — Part 1: Design, Fabrication, Inspection and Testing [178]. This standard includes requirements for the design, fabrication, inspection and testing of large transportable vacuum insulated cryogenic vessels with a volume of more than 450 litres, which are permanently (i.e. fixed tanks) or not permanently (i.e. demountable tanks and portable tanks) attached to a means of transport, for one or more modes of transport.
- ISO 20421-2:2017: Cryogenic Vessels — Large Transportable Vacuum Insulated Vessels — Part 2: Operational Requirements [179]. This standard includes operational requirements for large transportable vacuum insulated cryogenic vessels, including putting into service, filling, withdrawal, transport within the location, storage, maintenance, periodic inspection and emergency procedures.
- NFPA 55:2023: Compressed Gases and Cryogenic Fluids Code [180]. This standard covers protection from physiological, over-pressurization, explosive and flammability hazards associated with compressed gases and cryogenic fluids.

5.3.2. Hydrogen related best practices

More specific safety guidelines and recommendations have been developed in several safety related European Union projects. The following list is not exhaustive and is rather intended to illustrate the value and importance of pre-normative research.

- Pre-normative research on safe indoor use of fuel cells and hydrogen systems (HyIndoor[4]): This covers recommendations for safely using hydrogen in a closed space.
- Hydrogen Safety for Energy Applications (HySEA[5]): This covers information on safe venting of standard containers accommodating hydrogen technologies, as well as pre-normative research on vented deflagrations in enclosures and containers for hydrogen energy applications to introduce harmonized standard vent sizing requirements.

---

4 https://cordis.europa.eu/project/id/278534

5 www.hysea.eu

- Integrated Design for Efficient Advanced Liquefaction of Hydrogen (IDEALHY[6]): This covers safety assessment of liquid hydrogen production, transport and storage on a large scale.
- Prenormative Research for Safe Use of Liquid Hydrogen (PRESLHY[7]): This covers safety aspects of liquid hydrogen.
- Identification, Preparation and Dissemination of Hydrogen Safety Facts to Regulators and Public Safety Officials (HYFACTS[8]): This develops training material on hydrogen safety for public authorities.
- European Hydrogen Emergency Response training programme for First Responders (HyResponse[9]): This provides information for first responders.
- Support Safety Analysis of Hydrogen and Fuel Cell Technologies (SUSANA[10]): This establishes a framework for verification and validation for computational fluid dynamics to be used in risk assessment.

5.4. HYDROGEN SAFETY ASSESSMENT METHODOLOGIES

One of the most crucial factors for gaining government and public support for a nuclear hydrogen project is addressing and preparing for the possible safety risk involved. The safe use of hydrogen, especially on a large scale, calls for implementing an adequate risk management strategy, which calls for developing a robust safety assessment and risk analysis approach. To assess conformity to the requirements of RCSs, several methodologies are used. Table 18 (excerpt from Ref. [181]) presents methodologies commonly applied in industries using hydrogen. Hydrogen safety engineering uses validated tools and expertise from both nuclear and non-nuclear hydrogen safety fields to apply scientific and engineering principles to protect life, property and the environment from the adverse effects of incidents involving hydrogen.

---

6 https://www.idealhy.eu

7 https://preslhy.eu/

8 https://cordis.europa.eu/project/id/256823

9 https://cordis.europa.eu/project/id/325348

10 https://www.h2fc-net.eu/collaboration/susana-database/

TABLE 18. LIST OF HYDROGEN SAFETY ASSESSMENT METHODOLOGIES [181]

Methodology	Definition
Hazard identification	Hazard identification, or HAZID, is a systematic assessment to identify hazards and problem areas associated with plant, system, operation, design and maintenance. HAZID is used both as part of a quantitative risk assessment and as a standalone analysis for installation, modification, replacement and upgrade.
Hazard and operability analysis	Hazard and operability analysis, or HAZOP, is a qualitative method that systematically evaluates the causes, impact and consequences of deviations from expected or design values of the operating parameters, using project information. The method was developed to identify both hazards and operability problems at chemical process plants. Executing the method relies on using guidewords (e.g. 'no', 'more', 'less') combined with process parameters (e.g. temperature, flow, pressure) that aim to reveal deviations (e.g. less flow, more temperature) of the process intention or normal operation. The evaluation procedure is described in detail in IEC 61882:2016, Hazard and Operability Studies (HAZOP Studies) — Application Guide [182].
Failure modes and effects analysis	Failure modes and effects analysis is a tool to systematically analyse all contributing component failure modes and identify the resulting effect on the system. The semi-quantitative method is essentially composed of the following steps: — Define your system and the required level of analysis depth. — Identify hazards and events (potentially with an advanced HAZOP) for related equipment, components and processes. — Identify potential initiating failure modes and effects for all components and equipment and potentially early detection capabilities. — Determine a risk priority number. — Agree on acceptable limits for the risk priority number. — In case of limit violation, identify potential prevention and mitigation corrective action and re-evaluate risk priority number.
Risk matrix binning	Risk matrix binning is a qualitative method that combines the categorization of probabilities and consequences with risk acceptance categories in a matrix form. From this, management can distinguish whether a set of risks collected in one risk matrix cell has a higher priority than a set of risks collected in another cell.

TABLE 18. LIST OF HYDROGEN SAFETY ASSESSMENT METHODOLOGIES [181] (cont.)

Methodology	Definition
Probabilistic safety assessment	Probabilistic safety assessment uses probability distributions that encompass the range of possible values of variables in a risk equation to quantify the probability of the full range of potential outcomes. It is organized as a process for answering the following three questions: — What can go wrong? — How likely is it to happen? — What are the consequences? The main elements of a traditional probabilistic safety assessment are event tree and fault tree techniques, which constitute the logical model of the nuclear installation. Fault tree analysis is a quantitative (with regard to probabilities), deductive (or top–down) method used for the identification and analysis of conditions and factors that can result in the occurrence of a specific failure or undesirable event. This method addresses multiple failures, events and conditions. Event tree analysis is an inductive approach used for identifying and quantifying a set of possible outcomes, taking into account whether the installed safety barriers are functioning or not. The analysis starts with an initiating event or initial condition and includes the identification of a set of success and failure events that are combined to produce various outcomes. This quantitative method identifies the spectrum and severity of possible outcomes and determines their likelihood.
Quantitative risk assessment	Quantitative risk assessment is a formal and systematic method using measurable, objective data to determine an asset's value, the probability of loss and other associated risks. According to ISO Guide 73:2009, Risk Management — Vocabulary [183], risk is the combination of the probability of an event and its consequence. Explosion risk analysis is one of the most important components of the quantitative risk assessment and it could give accidental design loads that support the explosion risk-based mitigation design.
Structured 'what if' technique	The structured 'what if' technique, or SWIFT, is a speculative process where questions of the form 'What if (hardware, software, instrumentation, or operators) (fail, breach, break, lose functionality, reverse, etc.)?' are formulated and reviewed. SWIFT is sometimes described as a lightweight failure modes and effects analysis.

Note: Adapted in part from table A4-1 of Ref. [181]. Refer to the original table for the references relating to the descriptions.

IAEA Safety Standards Series No. GSR Part 4 (Rev. 1), Safety Assessment for Facilities and Activities [184], establishes the generally applicable requirements to be satisfied in safety assessment for facilities and activities that involve radiation, with a focus on defence in depth, safety margins, quantitative analyses and the use of a graded approach to the ranges of facilities and activities that are addressed. It provides details on the basis for requiring a safety assessment and the overall and specific requirements for the assessment. The deterministic safety analysis for nuclear power plants is addressed in IAEA Safety Standards Series No. SSG-2 (Rev. 1), Deterministic Safety Analysis for Nuclear Power Plants [185], while the development and application of Level 1 probabilistic safety assessment for nuclear power plants is covered in IAEA Safety Standards Series No. SSG-3 (Rev. 1), Development and Application of Level 1 Probabilistic Safety Assessment for Nuclear Power Plants [186], and Level 2 probabilistic safety assessment in IAEA Safety Standards Series No. SSG-4, Development and Application of Level 2 Probabilistic Safety Assessment for Nuclear Power Plants [187]. The results of the safety assessment are expected to be reflected in the safety analysis report, which needs to be in line with IAEA Safety Standards Series No. SSG-61, Format and Content of the Safety Analysis Report for Nuclear Power Plants [188].

5.5. COUPLING AND COLLOCATING NUCLEAR AND CHEMICAL FACILITIES FOR HYDROGEN PRODUCTION, STORAGE AND TRANSPORT

Collocation can be defined as the location and siting of a hydrogen facility inside the fence of a nuclear power plant, whereas coupling refers to locating or siting a hydrogen facility outside the fence of a nuclear power plant, with the distance from the perimeter depending on the accessibility of energy from the power plant without significant reduction in efficiency. These are covered in more detail in Section 3.3, which also reflects on the economic perspectives for both coupling and collocation.

The safety and regulatory aspects involve the technologies used as well as materials, typically:

- The nuclear power plant itself, when hydrogen is produced in a facility either coupled or collocated with the nuclear power plant;
- The hydrogen facilities (for the production, storage and transport) in general, with associated explosion risk;
- The associated chemical processes that may involve toxic substances and the potential of their release into the environment.

Several technologies are foreseen to be used in the production of nuclear hydrogen, such as:

- Steam methane reforming, which might use mainly high temperature heat and also electricity from a nuclear reactor;
- Electrolysis (low and high temperature);
- Thermochemical cycles involving chemical substances, some with high toxicity and high flammability risk;
- Pyrolysis, which might use mainly high temperature heat and also electricity from a nuclear reactor and produce solid carbon together with hydrogen.

These technologies, as well as their respective coupling with the nuclear power plant, are explained in more detail in Section 4 of this publication.

The specific siting requirements for collocating or coupling a hydrogen production facility with a nuclear power plant depend on the nuclear reactor type, the hydrogen production process, the distances between the hydrogen production facility and the residential areas in its vicinity, and prevailing public attitudes relating to acceptance [189].

Where hydrogen production systems are connected to a nuclear power plant, the following need to be considered: the hydrogen production system, as the final heat sink for the nuclear reactor, and the presence of flammable and toxic materials in the system and in the vicinity of the nuclear power plant. Potential hazardous events could be:

- Tritium migration from the reactor core to the hydrogen that is produced in the coupled plant;
- Thermal turbulence induced by failures or dysfunctionalities in the chemical system;
- Fire and explosion of flammable mixtures with process gases;
- Release of toxic material into the environment.

A review of Ref. [159] highlights the differences between collocation and coupling in terms of RCSs and requirements. Moreover, the existing RCSs identified were assessed to have limited applicability to collocating or coupling nuclear power plants with hydrogen production or storage facilities.

Because of the hazards associated with hydrogen production and storage, a risk assessment would be required to determine the reliability of the hydrogen production facility and to evaluate the consequences of potential accident scenarios posed by the hydrogen facility towards the nuclear power plants. On the other hand, the stability of the nuclear power plants' electrical and/or heat system and their effect on hydrogen production has to be assessed too. In fact, nuclear power plants are designed and licensed for a loss of load event. However, depending on the power distribution provided to the hydrogen production facility (dedicated or off grid) and the reliability and frequency of load disturbances, the licensee is likely to need to review the stability of the electrical system.

As the collocation of the nuclear power plants and hydrogen production and storage facilities may affect the emergency planning of both, a review of the existing emergency plans is needed to consider the effect that an accident or failure at the hydrogen production and storage facility may have on evacuation strategies in the event of a radiological emergency and/or possible toxic chemical release from the hydrogen facility. To minimize the consequences of accidents at both facilities (i.e. nuclear power plants and hydrogen production and storage), the safety measures discussed in the following paragraphs can be considered.

5.5.1. Separation distance

Only electricity needs to be supplied for LTE production facilities, and the most obvious approach is to provide sufficient separation so that the nuclear power plant cannot be influenced by events at the hydrogen production site. Concerning HTE production facilities, studies conducted by Idaho National Laboratory, the JAEA and Forschungszentrum Jülich [100] have determined that it is feasible to safely locate the hydrogen production and storage facilities at a minimum distance of 100 m from nuclear power plants without the need for mitigating barriers. A Chinese study proposes a distance of several hundred metres [190].

Table 19 provides examples of coupled LTE and HTE production facilities with nuclear power plants in the USA [159]. These projects are demonstrative, and they are described in more detail in Section 4 of this publication.

TABLE 19. EXAMPLES OF COUPLED NUCLEAR POWER PLANTS AND HYDROGEN PRODUCTION FACILITIES IN THE USA
(based on Ref. [159])

Station name and location	Hydrogen technology and size
Davis–Besse Nuclear Power Station, Toledo, OH	PEM, 1–3 MWe
Prairie Island Nuclear Generating Plant, Red Wing, MN	HTSE, 100–250 kWe
Palo Verde Nuclear Generating Station, Tonopah, AZ	PEM
Nine Mile Point Nuclear Station, Scriba, NY	PEM, 1.25 MWe

For steam methane reforming technologies, it is worth noting that existing large scale hydrogen production technologies are frequently located in urban areas. The approval of these industrial facilities suggests that they are safe to operate in urban areas and there is some degree of public acceptance. The operating experience from these industrial plants could provide confidence for regulatory approval. Nevertheless, their coupling with nuclear power plants needs to be assessed.

In all cases, the safety of separation distances has to be evaluated using state of the art tools and considering specific site commitments and licences, configurations, conditions and hazard and safety assessments.

5.5.2. Other safety measures

In addition to separation distance, precautions have to be taken to minimize the risk of a fire or gas explosion. Increased precautionary measures may be adopted for the reduction of combustibles in the vicinity of the nuclear power plant or the installation or strengthening of protection barriers. Additional design features may include redundant safety systems, physical separation by distance, separate fire compartments and fire protection systems. Moreover, ventilation systems in areas such as the nuclear power plant control room have to be designed in such a way that an ingress of flammable gases is prevented.

To overcome the limited applicability of RCSs to nuclear hydrogen production, pre-normative research is needed to fill the knowledge gap and enhance existing regulations and standards.

5.5.3. Integration of a hydrogen production facility with a nuclear power plant

Reference [100] includes a dedicated section on the safety of nuclear hydrogen production, covering integration of nuclear–chemical plants, basic considerations for a hydrogen safety system, tritium contamination aspects, explosion hazards, release of toxic materials and international regulations for handling hazardous materials.

In the nuclear industry, the safety standards and requirements for facilities and activities are provided in GSR Part 4 (Rev. 1) [184], which indicates that:

> "For sites with multiple facilities or multiple activities, account shall be taken in the safety assessment of the effects of external events on all facilities and activities, including the possibility of concurrent events affecting different facilities and activities, and of the potential hazards presented by each facility or activity to the others."

On nuclear power plants used for cogeneration of heat and power, Requirement 35 of IAEA Safety Standards Series No. SSR-2/1 (Rev. 1), Safety of Nuclear Power Plants: Design [191], stipulates that:

> **"Nuclear power plants coupled with heat utilization units (such as for district heating) and/or water desalination units shall be designed to prevent processes that transport radionuclides from the nuclear plant to the desalination unit or the district heating unit under conditions of operational states and in accident conditions."**

Although no mention of a hydrogen production plant is made under this requirement, the requirement is applicable also to hydrogen plants integrated with nuclear power plants.

In the USA, design changes are frequently carried out at operating nuclear power plants, in part by a procedure where the licensee validates that the proposed design change is allowed under the US Code of Federal Regulations 10 CFR 50.59 on changes, tests and experiments [192]. Regulations in 10 CFR 50.59 provide a threshold for determining when it is necessary to obtain approval of changes, tests or experiments from the Nuclear Regulatory Commission to preserve the basis on which the facility operating licence was issued. The intent of the 10 CFR 50.59 process is to permit licensees to make changes to the facility without needing an approval from the Nuclear Regulatory Commission through a licence amendment, provided that the changes maintain acceptable levels of safety as documented in the safety analysis report. A formal licence amendment request and a specific approval by the commission are required under 10 CFR 50.90 when a planned alteration to the facility is found to be outside the parameters outlined in 10 CFR 50.59. The licence amendment has to be obtained prior to implementing a proposed change, test or experiment if the change, test or experiment would (see Ref. [192]):

— "Result in more than a minimal increase in the frequency of occurrence of an accident previously evaluated in the final safety analysis report (as updated)". For example, steam diversion has the potential to impact safety analysis transients. As a follow up, assessment of the impact of turbine tripping frequency due to new transients and the evaluation of frequency increase of load/loss event have to be conducted.

— "Result in more than a minimal increase in the likelihood of occurrence of a malfunction of a structure, system, or component (SSC) important to safety previously evaluated in the final safety analysis report (as updated)". The evaluation of new pipe routing and location of the hydrogen production facility (and/or hydrogen storage, as applicable) and the impact on safety related equipment placed in proximity have to be assessed. For example, explosion of the hydrogen production facility or hydrogen storage tanks could cause nearby safety related equipment to malfunction or be fully damaged. Provision of adequate barriers and ensuring the safety distance between the hydrogen production facility and/or storage and equipment in proximity are mitigation measures to be taken.

- "Result in more than a minimal increase in the consequences of an accident previously evaluated in the final safety analysis report (as updated)". This involves the assessment of any radiological implication resulting from the change imposed by the coupling of the nuclear island with the hydrogen production facility.
- "Result in more than a minimal increase in the consequences of a malfunction of an SSC important to safety previously evaluated in the final safety analysis report (as updated)". For example, failure of the new steam piping could result in radiological consequences that will need to be evaluated.
- "Create a possibility for an accident of a different type than any previously evaluated in the final safety analysis report (as updated)". An assessment has to be conducted of the risk and impact of explosion and fire due to the hydrogen production plant and related equipment, including hydrogen storage, in connection with the nuclear power plant.
- "Create a possibility for a malfunction of an SSC important to safety with a different result than any previously evaluated in the final safety analysis report (as updated)".
- "Result in a design basis limit for a fission product barrier as described in the [final safety analysis report] FSAR (as updated) being exceeded or altered".
- "Result in a departure from a method of evaluation described in the [final safety analysis report] FSAR (as updated) used in establishing the design bases or in the safety analyses".

In order to reduce the uncertainty of the licence amendment request process, the Hydrogen Regulatory Research Review Group was established to identify the general technical and safety risks that may be acceptable under a 10 CFR 50.59 examination. This is expected to decrease the need for intricate regulatory clearances under the licence amendment request process. Ref. [193] reviews and discusses the results of regulatory research with a focus on the likelihood that the 10 CFR 50.59 utility self approval process will be employed. The generic 10 CFR 50.59 evaluation that results from these regulatory R&D based conclusions is included in Ref. [193]. A preliminary 10 CFR 50.59 evaluation for the addition of a collocated 100 MW high temperature electrolyser to a reference nuclear power plant (1200 MWe pressurized water reactor) has been prepared and is included in Ref. [193] to assist utilities in developing their own 10 CFR 50.59 evaluations for future hydrogen production facilities.

5.6. INTERNATIONAL EXPERTISE IN HYDROGEN SAFETY AND TARGETED PRE-NORMATIVE RESEARCH

The nuclear accident at the pressurized water reactor at Three Mile Island in 1979 prompted international research on severe accident phenomena and later the development of (severe) accident management programmes for nuclear power plants. Thus, intensive effort was made to better understand phenomena such as combustible gas generation, release, distribution, combustion and recombination in the containment of nuclear power plants [194]. In parallel, mitigation measures were developed and implemented in nuclear power plants. After the explosions occurred during the accidents at the Fukushima Daiichi nuclear power plant (Units 1–3), additional efforts were made to strengthen ways to manage the hydrogen risk [194]. The efforts also allowed for the development of tools and methodologies to assess the risks that are now used in design and licensing processes.

Developing RCSs for hydrogen production using nuclear reactors requires a holistic understanding of safety philosophies and approaches for both nuclear and hydrogen safety. To this end, global non-profit organizations such as the International Association for Hydrogen Safety (HySafe[11]) and the Center for Hydrogen Safety[12] provide knowledge and communication platforms for relevant hydrogen safety topics.

HySafe, based in Brussels, Belgium, is composed of approximately 50 member institutions from academia, research and industry with experiences in hydrogen safety. The association organizes Research Priority Workshops, the International Conference on Hydrogen Safety, and issues the Hydrogen Safety for Energy Applications handbook to continuously advance and disseminate the state of knowledge on hydrogen safety. The Center for Hydrogen Safety, which is based in the USA and has more than 100 member organizations, provides hydrogen safety information, guidance and expertise through the free Hydrogen Tools on-line portal[13], as well as webinars and conferences in collaboration with the American Institute of Chemical Engineers.

Canadian Nuclear Laboratories established the Canadian Hydrogen Safety Center[14] in 2023 to deliver tangible hydrogen safety solutions across multiple

[11] https://hysafe.info

[12] https://h2tools.org/

[13] https://www.aiche.org/chs

[14] https://www.ch2sc.ca/

industrial sectors and regions. It coordinates Canada's hydrogen safety expertise and capabilities and supports the development and implementation of RCSs.

5.7. GAPS IN REGULATIONS, CODES AND STANDARDS FOR HYDROGEN PRODUCTION PLANTS COUPLED OR COLLOCATED WITH NUCLEAR POWER PLANTS

To comply with the IAEA general safety requirements covered in GSR Part 4 (Rev. 1) [184], the safety assessment of hydrogen facilities coupled or collocated with nuclear power plants has to demonstrate to the satisfaction of the regulatory body that the proposed safety measures are adequate. Nonetheless, nuclear hydrogen production and storage implementation is dependent on the differences in RCSs used in the nuclear and chemical industries. Moreover, the existing RCSs identified have limited applicability to nuclear facilities used for hydrogen production. A detailed gap analysis on the RCSs for nuclear hydrogen production has to be completed by engaging multiple organizations, such as government regulatory bodies and industries.

The nuclear boundary has to be clearly established as this has an impact on the regulations and oversight of the hydrogen production facility by the nuclear regulator. Additionally, emergency planning considerations and strategies need to be established and codified. Clearly defined legal and regulatory responsibilities are needed (i.e. where and to what extent are the nuclear regulators or commercial/industrial regulators meant to be involved).

Targeted pre-normative research is needed to fill the following gaps [159]:

— Gaps and considerations relating to supply of feedstock:
  - Codes and standards and safety analyses for pipelines relating to supply of feedstock for hydrogen production within a nuclear perimeter (e.g. methane supply in the case of steam methane reforming);
  - Suitable codes and standards to accommodate the additional traffic required to transport hydrogen and pipelines entering and leaving the nuclear boundary;
  - Safety restrictions associated with flammable substances within a nuclear boundary;
  - Hazard analyses to prevent dust explosions, self heating and ignition, off-gassing and other hazards, such as toxic gas release;
  - Codes and standards relating to structural integrity and potential safety interferences caused by the supply of feedstock to the associated structures, systems and components of the combined nuclear power plant and hydrogen facility system;

  - Adequate quality assurance and control requirements and estimate of the feedstock required by the hydrogen production process.
- Gaps and considerations relating to hydrogen production:
  - Prompt and adequate response to transient states of the nuclear installations, sudden changes in electricity or steam supplies, as well as resilience of infrastructure to rapid pressure surges within the peripheral systems;
  - Prevention of the migration of tritium from the reactor core into the hydrogen produced;
  - Prevention of thermal turbulence induced by failures or dysfunctionalities in the chemical system.
- Gaps and considerations relating to storage management and safety:
  - Safe distancing between storage medium, nuclear power plant, hydrogen production system and peripheral systems, including impacts of accidental hydrogen release or explosion;
  - Codes and standards defining hydrogen storage and inventories allowed within the nuclear boundary;
  - Necessary response to detected problems and failures within the hydrogen storage and distribution system;
  - Possible auto-ignition of hydrogen (at 585°C), if leaked directly from a high temperature process (e.g. HTE, steam methane reforming).
- Gaps and considerations relating to hydrogen distribution:
  - Codes and standards and safety analyses for pipelines within a nuclear boundary.
  - Suitable materials for high pressure hydrogen pipelines, and for natural gas pipelines with higher concentrations of injected hydrogen; produced hydrogen (which could be distributed using existing natural gas pipelines, based on pipeline material and age, and the hydrogen concentration could vary between 4% and 30% by volume).
  - On-line measurement system to detect possible permeated hydrogen isotopes.
  - Codes and standards to determine an appropriate nuclear exclusion zone that are based on the size of the nuclear power plant and additional risk imposed by the hydrogen distribution network, if applicable. Large scale hydrogen production may require pipelines entering and leaving the nuclear boundary. Small scale hydrogen production may require trucks to transport hydrogen, resulting in additional traffic within the nuclear boundary.

In addition, the associated safety considerations derived from the existence of necessary feedstock and the emergence of by-products (such as oxygen,

CO/CO_2, and other chemicals, depending on the hydrogen production process) need to be incorporated in the RCSs that apply to nuclear hydrogen production.

6. STAKEHOLDER ENGAGEMENT

6.1. INTRODUCTION

Stakeholder engagement is an essential component in the development of any nuclear power programme. It is a strategic decision making tool that needs to be considered from the earliest stages of programme development and throughout the full life cycle, and it is achieved through open communication by the government, the regulatory body, the reactor owner/operator and other organizations and institutions of relevance for the programme. Stakeholder engagement is also key when nuclear reactor technology is used to support a larger project, such as the commercial deployment of hydrogen production.

As defined in Ref. [195], stakeholders can be 'statutory' (i.e. those required by law to be involved in any planning, development or operation of a nuclear project) or 'non-statutory' (i.e. those who have an interest in or will be directly or indirectly impacted). Both categories will need to be considered in the lifetime of a nuclear hydrogen project. Reference [195] provides details on the importance of stakeholder engagement, its key principles, the methodology for developing stakeholder engagement in a nuclear power programme, and the relevant roles and responsibilities of key organizations, all of which are widely applicable to a nuclear hydrogen project as well. It is important to note that a stakeholder is inherently "any group or individual who feels affected by an activity, whether physically or emotionally" [195].

Setting up a stakeholder engagement programme for nuclear hydrogen production includes developing strategic plans for engagement, aligning the engagement strategy with all milestones in the project roadmap, mobilizing sufficient and suitable human and financial resources, identifying and mapping stakeholders and crafting key messages to support interaction. Table 20, adapted from Ref. [195], provides insights into formulating a comprehensive stakeholder engagement strategy for nuclear hydrogen deployment and includes guiding questions to support such strategy development. While it is possible and encouraged to base engagement on other nuclear activities and facilities, such as nuclear power plants and radioactive waste management facilities, it is important to carefully consider the unique characteristics of nuclear hydrogen and their implication in terms of stakeholder engagement.

Researching the interests, opinions, knowledge and feelings of different stakeholders with regards to nuclear hydrogen is an important component for supporting project development. The research can be undertaken in different ways, including public surveys, focus groups, executive interviews and two-way communication with stakeholders. The results of this research will help to inform relevant stakeholder engagement activities.

Defining clear roles and responsibilities of the key organizations involved in the deployment of a nuclear hydrogen project provides a foundation for strategic stakeholder engagement. The government, regulatory body and the operating organization will each develop their respective stakeholder engagement programmes that include their interaction with each other.

TABLE 20. GUIDING QUESTIONS FOR DRAFTING A STAKEHOLDER ENGAGEMENT STRATEGY FOR NUCLEAR HYDROGEN DEPLOYMENT [195]

Strategy components	Guiding questions
Introduction, including background/situation analysis	
Describe the energy and socioeconomic situation of the country and the national strategy for the deployment of nuclear hydrogen (e.g. as it relates to energy security, decarbonization, clean energy, other factors). Describe related stakeholder engagement activities to date. Include a situational analysis of the current versus the desired situation.	— Why is nuclear hydrogen deployment being considered?
Objectives	
Briefly provide the purpose of the stakeholder engagement strategy in a few short paragraphs that outline what the strategy will aim to accomplish.	— What has led to this engagement activity? — Is there an opportunity for stakeholders to influence decisions, policy or the project? — What knowledge do you have that you can build on? — Is everyone clear about the decisions to be made? — Does the purpose statement reflect the needs of the decision makers and stakeholders? — What is the purpose for communicating key programme messages?

TABLE 20. GUIDING QUESTIONS FOR DRAFTING A STAKEHOLDER ENGAGEMENT STRATEGY FOR NUCLEAR HYDROGEN DEPLOYMENT [195] (cont.)

Strategy components	Guiding questions
Scope	
Describe what the strategy covers (e.g. stakeholder engagement activities relating to nuclear hydrogen, including public information, coordination among key organizations, engaging with other identified stakeholders) and what it does not cover.	— What aspects of the overall nuclear hydrogen production programme will be separate from this stakeholder engagement strategy?
Key stakeholders	
List the stakeholders who have been identified (statutory and non-statutory) and include some form of prioritization based on their levels of interest in and decision making power for the nuclear hydrogen production programme (be as specific as possible).	— Will these stakeholders help to achieve your objectives? — Are you involving them because you need input to decide how to implement the plan? — Is there a high level of community impact? — Is the programme politically sensitive? — Who are you aware of that has: A direct and clear interest? A general interest? An ability to influence? — Who may not be interested but will be affected by the decision? — Which geographic areas do you want messages to reach? (Are there any areas that are not to receive the messages?) — Who needs to know details about your programme?

TABLE 20. GUIDING QUESTIONS FOR DRAFTING A STAKEHOLDER ENGAGEMENT STRATEGY FOR NUCLEAR HYDROGEN DEPLOYMENT [195] (cont.)

Strategy components	Guiding questions
High level messages	
List up to five key messages relating to the nuclear hydrogen production programme.	— What are the three to five key messages you want stakeholders to understand about the nuclear programme in order for the programme to succeed? — What questions do people ask staff during programmes/events? — What would you want to know about the programme if you lived in the community? — Is there any information that is not to be shared, given the context?
Timeline	
Identify the major steps in the nuclear hydrogen production programme, for example using a roadmap approach as described in Section 7 of this publication, to which the stakeholder engagement strategy needs to be closely aligned.	— What are the key activities in your programme? Which ones need to be preceded by information sharing? — Does the calendar of events allow for impromptu changes? — When will you share key information or updates with the different audiences?
Resources	
Briefly describe the necessary adequate human and financial resources for stakeholder engagement and anticipated expansion of resource needs as nuclear hydrogen deployment moves through the outlined time frame.	— What is the total budget required in order to implement the plan? — Is it clear who will have responsibility for delivery of each communication activity? — Do you have sufficient staff to deliver this strategy, or will outside resources be required? — What funding is already in place?

TABLE 20. GUIDING QUESTIONS FOR DRAFTING A STAKEHOLDER ENGAGEMENT STRATEGY FOR NUCLEAR HYDROGEN DEPLOYMENT [195] (cont.)

Strategy components	Guiding questions
Tools, methods and approaches	
Provide general information on the communication channels and tools to be used in order to deliver messages to relevant stakeholders.	— What channels will you use to communicate the messages to the target audience? — Do the selected channels pose any challenges to staff in terms of creation? — Do the selected channels pose any challenges to stakeholders in terms of accessibility?
Evaluation criteria and follow-up process	
Describe how the effectiveness of the implementation of the stakeholder engagement strategy will be evaluated, measured and adapted (e.g. through public opinion surveys, media, social media monitoring).	— Who will be responsible for developing the review criteria and making the review happen? — What methods will you use to decide whether each communication approach is effective?
References	
List documents that can provide necessary background and technical information.	— What are the relevant reference documents for the strategy?
Abbreviations	
Define abbreviations used in the strategy.	— What abbreviations are used in the strategy and are to be included as a reference for clarity to readers?
Distribution list	
Distribute the strategy to all individuals and organizations responsible for and/or participating in stakeholder engagement.	— Who needs to receive this strategy in order to carry out their work?

The following sections provide details on the salient stakeholder engagement considerations specific to nuclear hydrogen projects.

6.2. STAKEHOLDER ENGAGEMENT IN NUCLEAR HYDROGEN PROJECTS

The nature, interest and influence levels of stakeholders of a nuclear hydrogen project are expected to evolve in a dynamic way on the basis of a number of factors, including the expansion and development of the low carbon hydrogen market. The range of potential stakeholders for a hydrogen market is not completely known. As of 2019, only 5% of the produced hydrogen was traded [196], but this percentage is foreseen to increase, impacting imports and exports and thereby the associated stakeholders at both national and international levels.

Several countries have issued or are developing national hydrogen roadmaps, while energy policies are being shaped to include hydrogen in the clean energy transition planning, particularly as a route to industrial decarbonization. In this context, the government, nuclear power owners/operators, hydrogen producers (owners/operators of a hydrogen production facility), distribution infrastructure developers and the potential end users of hydrogen are relevant stakeholder groups of a nuclear hydrogen project. In addition, stakeholders associated with a nuclear power programme can be considered as stakeholders of a nuclear hydrogen project. In planning for nuclear hydrogen projects, mapping of the stakeholders is needed at an early stage to ensure that they are effectively engaged and that those interested in, and with a potential influence on, a project are included in the decision making process. Stakeholders can be at local, national and international levels, with their respective impacts on the dynamics and development of nuclear hydrogen projects, contributing experience from different demonstration projects and sharing relevant knowledge.

Insights into stakeholder perceptions and perceived challenges in relation to the development of a hydrogen economy in general can be found in various references. Key challenges and stakeholder mapping methodology, which allows users to estimate and compare the various prospective hydrogen supply chains, are covered in Refs [197, 198]. This methodology aims at increasing the qualitative information analysis value in hydrogen roadmapping activities. In Ref. [199], the current status of the energy and distribution system in Poland is reviewed. Policies for planning a transition to a hydrogen based economy that includes nuclear power and an analysis of them from short, medium and long term perspectives are also provided. A stakeholder analysis for hydrogen research in Denmark is given in Ref. [200], showing the importance of critical stakeholder analysis to identify the type of actors (e.g. corporate stakeholders, universities, institutions, utilities, industries) that can facilitate the technical and economic development of hydrogen and fuel cell technologies. In Ref. [201], the results of a survey on acceptance, an analysis of German news and media and an analysis of communication material to identify arguments used for and against hydrogen are provided. One of the key challenges identified is that the essential advantages of hydrogen as an energy carrier are rarely highlighted in communication materials and the media. A stakeholder analysis for Germany, with a focus on the period needed for hydrogen market ramp-up, is provided in Ref. [202]. The risks are identified and the potential roles of the economic, ecological and political sectors are suggested.

Section 3 of this publication illustrates the current and possible future hydrogen market. Markets are composed of different value chains. These value chains include a series of activities performed by various stakeholders involved in providing a specific good or service.

Value chains do not exist 'in a vacuum' but are shaped by an ecosystem that also includes other relevant stakeholders such as:

- Institutions (e.g. governments (which influence value chains through policies), regulators, financers);
- External stakeholders (e.g. local communities, non-governmental organizations);
- Technology suppliers.

Figure 25 presents the value chain and ecosystem for the case of nuclear hydrogen for sea shipping.

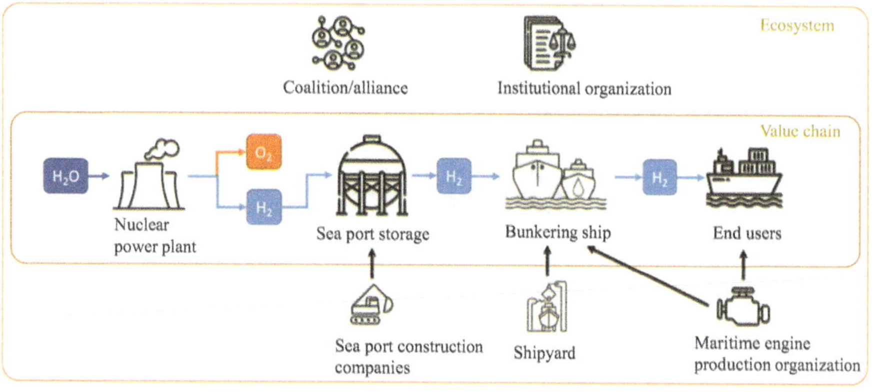

FIG. 25. Nuclear hydrogen for the sea shipping value chain and ecosystem.

Figure 26 shows only a few of the possible elements belonging to a generic nuclear value chain ecosystem. Stakeholders in the value chain include nuclear power plant operators, seaport storage operators, bunkering ship operators and organizations operating cargo ships to provide transport services. All of these stakeholders are directly linked with hydrogen production, storage, transport and consumption. The generic ecosystem includes the stakeholders within the value chain and other external stakeholders. For the case of sea shipping, the external stakeholders might be seaport construction companies (which provide hydrogen storage construction services), shipyards (which build hydrogen fuelled ships), marine engine production companies (which provide hydrogen propulsion systems), institutions (which standardize and regulate hydrogen applications), and coalitions or alliances (which are clusters of hydrogen specialized professionals and organizations). Even if such external stakeholders do not belong to the value chain, they can play key roles in setting up, shaping or supporting this value chain.

Engagement between stakeholders is, therefore, essential both at the value chain and ecosystem levels. Management literature shows that project promotion can be carried out in different ways, including through intermediaries. Universities, private companies and industrial organizations could play a fundamental role, as intermediaries, in shaping the value chain and ecosystems by promoting and supporting the deployment of nuclear hydrogen production.

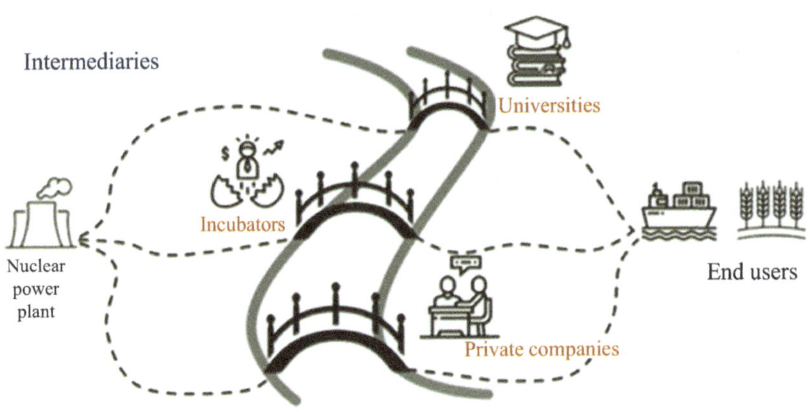

FIG. 26. Elements in nuclear hydrogen value chain ecosystems.

6.3. STAKEHOLDER IDENTIFICATION AND ENGAGEMENT IN NUCLEAR HYDROGEN PROJECTS

When identifying and mapping stakeholders of relevance for a nuclear hydrogen project, it can be helpful to consider those that are specific to the nuclear energy component of the project, those that are specific to the hydrogen part, and those that are common to both (see Fig. 27). According to Ref. [202], among the relevant stakeholders for a developing hydrogen market, R&D organizations play a critical role, followed by hydrogen technology providers, electricity utilities and others (such as water utilities and gas distribution infrastructure providers). These groups also have considerable contact with other stakeholder groups and, in the case of a nuclear hydrogen project, they are expected to interact to a significant extent with the stakeholders in the nuclear group. In the case of deploying a nuclear hydrogen project, some stakeholders might enter into competition or possible conflict, for example natural gas companies or renewable energy companies might compete with nuclear utilities for the hydrogen production/supply infrastructure.

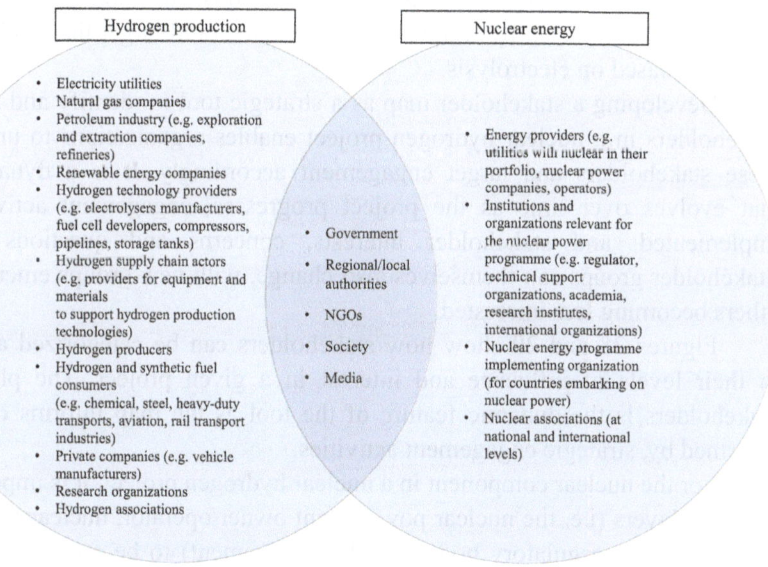

FIG. 27. Possible stakeholders of relevance for a nuclear hydrogen project.

The issue of direct electrification versus hydrogen utilization may also be perceived as a potential conflict, as in the case of domestic heating (e.g. heat pumps versus hydrogen boilers) or short distance transportation (e.g. battery electric vehicle versus fuel cell electric vehicle) [202]. Also, competition is expected to become more intense between hydrogen end users as new markets emerge and government, regional or municipal authorities prioritize and provide incentives and stimulation for different clean hydrogen production options. Since, in many countries, current policies favour the development of a low carbon hydrogen market to address climate goals, there are clearly opportunities for investments in new facilities for hydrogen production using low carbon energy sources, such as nuclear energy, and using more efficient equipment and technologies. At the same time, these opportunities would also involve some investment risks to be undertaken by the first movers that would test and demonstrate the new production technologies, processes and equipment. These risks may arise owing to the uncertainty in the evolution of the hydrogen market and in the cost of hydrogen production using different energy sources, technical challenges imposed by the different processes (in the production stage but also in the storage and transportation stages) or regulatory challenges that are expected when the regulatory framework has not been established. Another risk may arise for the regions affected by water scarcity, which would target the production of hydrogen based on electrolysis.

Developing a stakeholder map as a strategic tool to identify and prioritize stakeholders in a nuclear hydrogen project enables organizations to understand these stakeholders and target engagement accordingly. It is a dynamic tool that evolves over time as the project progresses, engagement activities are implemented, and stakeholder interests, concerns and questions change. Stakeholder groups can themselves also change, with new groups emerging and others becoming less interested.

Figures 28 and 29 show how stakeholders can be categorized according to their levels of influence and interest in a given project. The plotting of stakeholders is the dynamic feature of the tool as the map informs of, and is informed by, strategic engagement activities.

For the nuclear component in a nuclear hydrogen project, it is important for the key players (i.e. the nuclear power plant owner/operator, nuclear technology developers, the regulatory body and the government) to be aware of potential barriers and uncertainties with regards to stakeholder engagement, some of which are identified in Table 21.

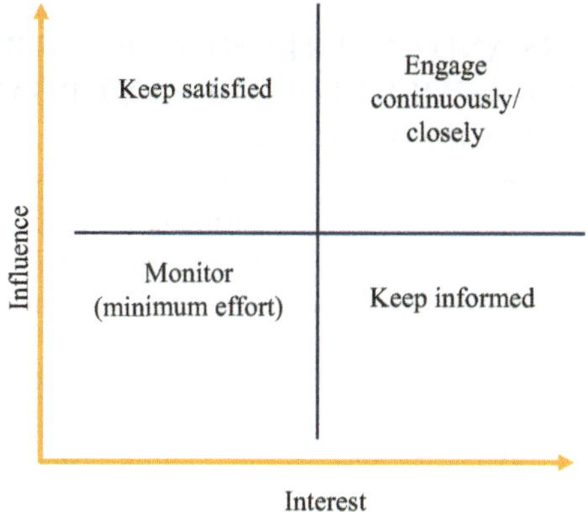

FIG. 28. Example of a stakeholder mapping matrix.

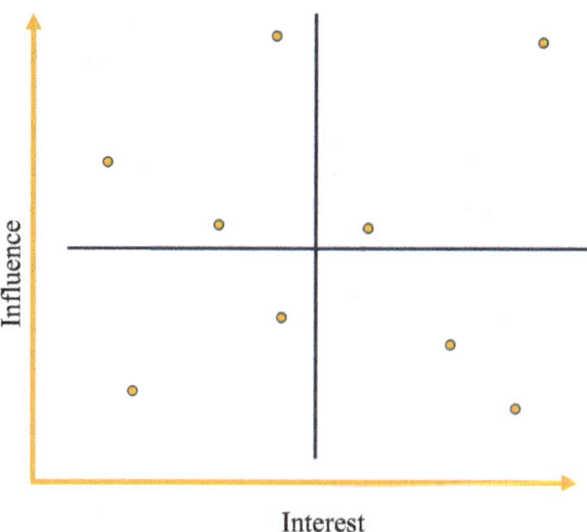

FIG. 29. Illustration of groups on a completed stakeholder map.

TABLE 21. BARRIERS AND UNCERTAINTIES RELATING TO STAKEHOLDER ENGAGEMENT FOR THE KEY ORGANIZATIONS IN A NUCLEAR HYDROGEN PROJECT

Key player	Possible barriers and uncertainties
NPP operator	— Ageing fleet: Licence renewal when/if investments are needed for component replacements. — Operating fleet: Engineering modifications may be necessary for a nuclear hydrogen project, requiring time and financial resources as well as possible regulatory engagement.
Nuclear technology developers	Uncertainties associated with the target audience and further market development: — Understanding who the target customer is and how those customers define the 'value' of nuclear hydrogen. — The shape and size of the market in 5–10 years is barely known, which might cause uncertainties and concerns for project developers and investors. — The funding needs, sources and their availability need to be carefully considered, as significant capital costs may occur. This might cause uncertainties and concerns for project developers and investors.
Nuclear regulatory body	— The extent to which the non-nuclear regulations are applicable and the extent to which new regulations and frameworks have to be prepared for a nuclear hydrogen project. This has to be clarified before the initiation of a nuclear hydrogen project. — The regulator needs to understand the boundaries of its assessment and licensing (in the case of a nuclear hydrogen project): Would its expertise and licence be extended to the hydrogen production facility? What are the regulatory modifications needed for an NPP coupled with a hydrogen production facility? Engagement by the nuclear regulator with the government, the nuclear facility owner/operator and the regulator of the hydrogen production facility is needed. — Involvement of non-nuclear regulatory bodies (e.g. agencies associated with licensing of chemical plants where bulk hydrogen production and handling are already being practised) may be needed to ensure adequate technical competence of the regulatory authorities that may otherwise be unfamiliar with hydrogen specific considerations.

TABLE 21. BARRIERS AND UNCERTAINTIES RELATING TO STAKEHOLDER ENGAGEMENT FOR THE KEY ORGANIZATIONS IN A NUCLEAR HYDROGEN PROJECT (cont.)

Key player	Possible barriers and uncertainties
Nuclear regulatory body	— How much interest currently exists regarding deployment of nuclear hydrogen projects (e.g. existing or new reactor designs, any expected plant modifications, possibility of optimization for nuclear hydrogen production), and from which NPP operators/ reactor technology developers? — The staff needed to complete reviews and licensing might not be fully familiar with the nuclear hydrogen technologies.
Government	The government may have the following concerns about the distribution of possible risks in relation to a nuclear hydrogen project: — Who assumes the risk (public/private/others)? — Who benefits from nuclear hydrogen implementation? — Does the government support hydrogen subsidies in general (whether nuclear or not)? — What is the timing of investment return (if applicable)? — Is there consensus/divergence between political parties? — Relationship with the public, how to ensure public confidence or trust?

Note: NPP — nuclear power plant.

6.4. KEY CONSIDERATIONS FOR STAKEHOLDER ENGAGEMENT IN NUCLEAR HYDROGEN DEPLOYMENT

To accelerate nuclear hydrogen deployment, the following steps need to be considered at a national level when addressing different groups of stakeholders:

- Engage nuclear power plant owners/operators in discussions to clearly elaborate the roles for nuclear hydrogen as a commodity (reliable supply) instead of just as an energy vector (flexibility).
- Work on fostering public understanding of the contribution of hydrogen production for energy decarbonization towards achieving climate goals and the benefits of a partnership between renewable and nuclear fuels in an integrated decarbonized grid. The public may question the necessity considering the availability of alternative fuel that is currently cheaper and with a clearer principle of use. Engagement efforts can emphasize the need

for baseload electricity supply, as well as the benefits of nuclear energy in a diverse, integrated, reliable and low carbon energy system.

— While underscoring the benefits in a clear and understandable way is an important component for stakeholder engagement efforts by the nuclear hydrogen project developer, the overarching strategy needs to include mechanisms for addressing safety and other concerns, as applicable, in a transparent and open manner.

For the deployment of nuclear hydrogen, many elements of demonstrated stakeholder engagement practices in the nuclear power sector can also be applied. A strategic approach to stakeholder engagement as presented in this section can support project progress and success.

7. ROADMAP FOR THE COMMERCIAL DEPLOYMENT OF NUCLEAR HYDROGEN PRODUCTION

7.1. CONCEPT AND FEATURES OF ROADMAPS

A roadmap is a strategic tool often used for planning and coordination functions to support management decisions. Roadmaps are very often created and used by high tech industries and corporate organizations, including sectors such as electronics, software and information technology, aviation, space and automotive. They are prepared in the context of critical technological need gap assessments, new product or technology development and decisions relating to its commercial deployment, formulation of associated policies and establishment of financing or investment strategies. They also often include the activity timelines and milestones to be achieved on the path from conceptualization of the project to its successful implementation. They can take many visual formats (often graphical depictions with brief textual content) and serve multiple purposes, depending on the industry or sector concerned, the specific type or nature of the project, and the potential uses and users of the roadmap.

The various forms of roadmaps and examples of their usage are well described in technology management literature, such as Ref. [203]. An example of a generic technology roadmap is provided in Fig. 30, which shows various activities, and subactivities to be performed under different workstreams, from initial conceptualization to completion of the project. Once drawn up, the roadmap is typically used as the basis to chalk out the detailed action plan, from concept to commissioning.

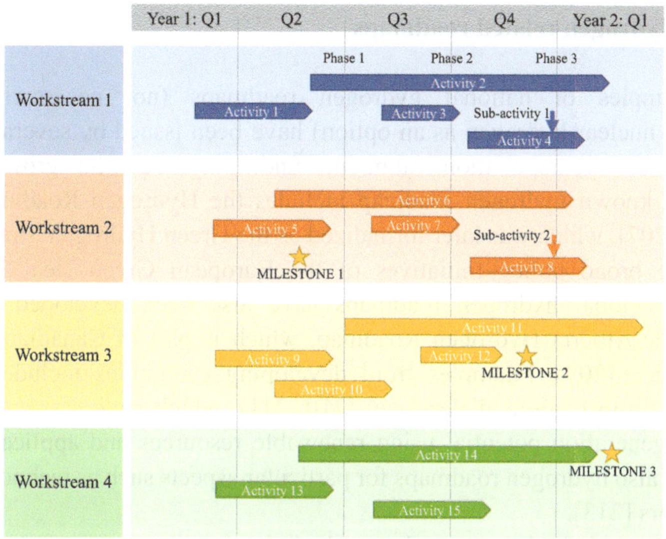

FIG. 30. Example of a generic technology roadmap showing individual workstreams and activities under specific phases, from initial conceptualization to the completion of the project.

7.2. EXAMPLES OF ROADMAPS IN THE NUCLEAR AND HYDROGEN SECTORS

7.2.1. Nuclear power related roadmaps

Road mapping exercises have been an important part of the assessments performed by a country when considering deployment or extension of its nuclear power programme. Under the International Project on Innovative Nuclear Reactors and Fuel Cycles (INPRO) launched by the IAEA in 2000, the methodology for the assessment of multidimensional sustainability characteristics of nuclear energy programmes was developed and extended to a collaborative project on roadmap development for nuclear power deployment [204].

The IAEA has also issued a publication on roadmaps for the deployment of SMRs, addressing the series of initiatives, preparatory activities and milestones to be achieved by potential reactor owners/operators and technology developers and vendors for implementing a country's first SMR project [205]. These roadmaps are also closely aligned with the IAEA Milestones approach [206], which refers to 19 infrastructural issues and associated milestones that have to be adequately taken care of by a country wishing to deploy and/or extend a civilian nuclear power programme. Country specific roadmaps for advanced nuclear technologies are also available (e.g. the Canada SMR roadmap [189]).

7.2.2. Hydrogen related roadmaps

Examples of national hydrogen roadmaps (not necessarily always including nuclear hydrogen as an option) have been issued by several countries in the last few years, as momentum surrounding hydrogen has grown globally. A widely known hydrogen roadmap includes the Hydrogen Roadmap Europe of 2016 [207], which was later formalized as the Green Hydrogen Strategy [208] under the broad policy initiatives of the European Green Deal[15]. Within a country, regional hydrogen roadmaps have also been developed, as in the case of the Alberta Hydrogen Roadmap, which is part of Canadian initiatives in hydrogen [209]. Examples from developing economies include hydrogen roadmaps from Chile and Paraguay [210, 211], which have a strong focus on regional generation potential using renewable resources and application areas. There are also hydrogen roadmaps for particular aspects such as technology [212] or investors [213].

In general, hydrogen roadmaps include an assessment of the hydrogen needs, hydrogen production possibilities using domestic resources (e.g. energy, land, water, technical know-how), need for imports or potential for exports of hydrogen or hydrogen derivatives of the region or country concerned (including mechanisms and timeframes for bilateral agreements and deals relating to hydrogen or its derivatives), the costs (e.g. the levelized costs of hydrogen production, storage and distribution) and benefits associated with these projects (e.g. extent of decarbonization, emissions reduction, reduction of energy imports, new employment opportunities), gap areas for R&D, funding and financing opportunities, and a set of recommendations for policy action.

A nuclear hydrogen roadmap has therefore to be designed along similar lines, while specifically addressing the unique considerations associated with the coupling of hydrogen plants with nuclear plants (e.g. whether the facility is collocated with the nuclear power plant or otherwise, potential safety impacts of the hydrogen production plant on the operations of the nuclear reactor). In 2021, a review of governments with adopted national hydrogen strategies, announced targets, priorities for hydrogen use, and committed funding for hydrogen projects was issued by the IEA [51]. Table 22, adapted from Ref. [51], presents Member States considering nuclear hydrogen production within their national hydrogen strategies.

[15] Further information on the European Green Deal can be found at: https://commission.europa.eu/strategy-and-policy/priorities-2019-2024/european-green-deal_en

TABLE 22. MEMBER STATES CONSIDERING NUCLEAR HYDROGEN PRODUCTION WITHIN THEIR NATIONAL HYDROGEN STRATEGIES

(adapted from Ref. [51])

Member State	Reference/Role	Deployment targets (2030)	Production	Priority uses	Committed public investment
Canada	National Hydrogen Strategy, 2020 [214]	Total use: 4 Mt H_2/a, 6.2% total final energy consumption	Biomass; By-product; Electrolysis; Natural gas/oil with CCUS	Construction, Electricity, Exports, Industry, Mining, Refining, Shipping, Transport	~US $19 million by 2026
	Role of nuclear energy to support the hydrogen roadmap	Hydrogen production methods considered are: — Electrolysis using off-peak nuclear electricity in the near term; — High temperature thermal processes by coupling with SMRs, which are viable in the longer term. There are opportunities for nuclear hydrogen production in Ontario and in New Brunswick, and longer term in western and central Canada.			
Czech Republic	National Hydrogen Strategy, 2021 [215]	Low carbon demand: 97 kt H_2/a	Electrolysis	Industry (chemicals), Transport	—[a]
	Role of nuclear energy to support the hydrogen roadmap	Potential technologies to produce hydrogen in the Czech Republic are electrolysis with energy supplied by an NPP or the pyrolytic decomposition of natural gas with the processing/storage of the resulting carbon. SMRs coupled with electrolysers might be considered for hydrogen production in the future.			

TABLE 22. MEMBER STATES CONSIDERING NUCLEAR HYDROGEN PRODUCTION WITHIN THEIR NATIONAL HYDROGEN STRATEGIES (cont.)

(adapted from Ref. [51])

Member State	Reference/Role	Deployment targets (2030)	Production	Priority uses	Committed public investment
France	Hydrogen Deployment Plan, 2018 [216]; National Hydrogen Strategy, 2023 [217]	6.5 GW electrolysis, 20–40% industrial decarbonized hydrogen Electrolysis		Industry, Refining, Transport	~ US $8.2 billion by 2030
	Role of nuclear energy to support the hydrogen roadmap	While not explicitly stated in Refs [216, 217], France is considering using its nuclear fleet and SMRs to support hydrogen production [218]. It is also supporting the inclusion of nuclear hydrogen production under the low carbon classification in European legislation.			
Hungary	National Hydrogen Strategy, 2021 [219]	Production: 20,000 t/a of low carbon H_2, 16 kt/a of carbon free H_2, 240 MW electrolysis	Electrolysis; Fossil fuels with CCUS	Electricity, Industry, Transport	—
	Role of nuclear energy to support the hydrogen roadmap	The Paks NPP may supply a significant amount of carbon free electricity for the establishment of the hydrogen value chain.			

TABLE 22. MEMBER STATES CONSIDERING NUCLEAR HYDROGEN PRODUCTION WITHIN THEIR NATIONAL HYDROGEN STRATEGIES (cont.)

(adapted from Ref. [51])

Member State	Reference/Role	Deployment targets (2030)	Production	Priority uses	Committed public investment
Japan	Strategic Roadmap for Hydrogen and Fuel Cells, 2019 [220]; Green Growth Strategy, 2021 [221]	Total use: 3 Mt H_2/a Supply: 420 kt low carbon H_2	Electrolysis; Fossil fuels with CCUS	Construction, Electricity, Industry (steel), Refining, Shipping, Transport	~ US $6.5 billion by 2030
	Role of nuclear energy to support the hydrogen roadmap	Japan developed a roadmap of growth strategies for nuclear industry. It aims to establish carbon free hydrogen production technology utilizing high temperature heat coming from nuclear reactors (for hydrogen production using the I–S process, methane pyrolysis method and/or steam methane reforming) by 2030.			
Russian Federation	Hydrogen Roadmap, 2020 [222]	Exports: 2 Mt H_2	Electrolysis; Natural gas with CCUS	Electricity, Industry, Refining, Exports	—
	Role of nuclear energy to support the hydrogen roadmap	The Russian Federation is seeking to deploy a pilot project for hydrogen production using the capacities of the Russian NPPs. It also aims to develop the concept for ensuring safety in the production, storage and transportation of hydrogen at NPPs.			

TABLE 22. MEMBER STATES CONSIDERING NUCLEAR HYDROGEN PRODUCTION WITHIN THEIR NATIONAL HYDROGEN STRATEGIES (cont.)

(adapted from Ref. [51])

Member State	Reference/Role	Deployment targets (2030)	Production	Priority uses	Committed public investment
United Kingdom	UK Hydrogen Strategy, 2021 [223]	5 GW low carbon production capacity	Natural gas with CCUS; Electrolysis	Aviation, Buildings, Electricity, Industry, Refining, Shipping, Transport	~ US $1.3 billion
	Role of nuclear energy to support the hydrogen roadmap	Nuclear hydrogen production is expected by the mid-2030s and the following production methods are envisaged: — Low temperature nuclear electrolysis (can apply existing technologies to current NPPs in the 2020s); — High temperature nuclear electrolysis (could consider advanced nuclear reactors in the 2030s); — Thermochemical water splitting (could consider advanced nuclear reactors in the 2030s).			

TABLE 22. MEMBER STATES CONSIDERING NUCLEAR HYDROGEN PRODUCTION WITHIN THEIR NATIONAL HYDROGEN STRATEGIES (cont.)

(adapted from Ref. [51])

Member State	Reference/Role	Deployment targets (2030)	Production	Priority uses	Committed public investment
United States of America	DOE National Clean Hydrogen Strategy and Roadmap, 2023 [88]	10 Mt of clean hydrogen annually by 2030	Electrolysis using energy from nuclear reactors and renewables; Steam methane reforming with CCS	Back-up power, Aviation, Buildings, Electricity, Industry, Refining, Shipping, Transport	US $9.5 billion
	Role of nuclear energy to support the hydrogen roadmap	The DOE's H2Hubs program invests in hydrogen hubs across the USA to create a national network of hydrogen producers, consumers and infrastructure. These hubs aim to accelerate the commercial-scale deployment of clean hydrogen, helping to decarbonize heavy industry and transportation. The Nine Mile Point facility is the first-of-its-kind in the USA to generate clean hydrogen (560 kg/d) using Constellation's nuclear power plant in Oswego, New York. Constellation has pledged to invest $900 million through 2025 for commercial clean hydrogen production using nuclear energy as part of its larger decarbonization strategy. Constellation is currently pursuing the development of regional hydrogen production and distribution hubs in collaboration with public and private entities representing every stage of the hydrogen value chain. This includes involvement in the Northeast Clean Hydrogen Hub, Mid-Atlantic Hydrogen Hub, and Midwest Alliance for Clean Hydrogen (MachH2), all of which are looking into projects to build hydrogen infrastructure in partnership with DOE [224].			

Note: NPP — nuclear power plant; DOE — US Department of Energy.

[a] —: data not available.

7.3. CONSIDERATIONS FOR A NUCLEAR HYDROGEN ROADMAP

The need to develop a roadmap for the deployment of nuclear hydrogen production lies in understanding the rationale or justification behind creating such a programme, the various stakeholders likely to be involved, the expected time frame for the realization of such programmes and the associated costs and benefits from such programmes. The rationale and set of enabling factors for nuclear hydrogen deployment that a country can make use of may be indicated as follows [71]:

- Long term government commitment and support for nuclear power programmes in general, including understanding the role of non-electric applications of nuclear power;
- Industrial and/or user consortium initiatives between project developers and offtake industries;
- Development of green financial taxonomies and ensuring applicability of the taxonomy to all low carbon energy forms, including nuclear power;
- Adoption of a scientifically sound and technically definable approach to clean, or low carbon hydrogen, utilizing harmonized and verifiable life cycle emissions and carbon intensity accounting in accordance with international norms;
- Clearly identified role of nuclear hydrogen in national hydrogen missions or strategies and climate action plans;
- Making available benefits of market based mechanisms (e.g. carbon pricing, removal of fossil fuel subsidies) for funding or financing of nuclear hydrogen projects alongside other hydrogen sources;
- Regional, bilateral and multilateral resources and initiatives (e.g. those by supranational organizations such as the Generation IV International Forum, IAEA, OECD/NEA);
- Private–public (government) partnerships for technology supply, financing, offtake of produced hydrogen.

In its most general form, the nuclear hydrogen production roadmap has to address (among other things) the aspects described in Table 23. The activities and their sequence have to be planned accordingly, in consultation with the different stakeholders concerned.

TABLE 23. PRINCIPAL COMPONENTS OF A NUCLEAR HYDROGEN ROADMAP

Technology features	— Available hydrogen production technologies — Fresh water demand for the hydrogen plant and purification technologies — Technology and market readiness assessment of the hydrogen technologies — Technology suppliers — Plant size, performance and quality acceptance criteria — Methods and infrastructure for dispatch of hydrogen to users — Methods and infrastructure for managing by-products
Safety and regulatory aspects	— Type of coupling scheme between the hydrogen plant and the nuclear power plant — Potential impact on the safety of the nuclear power plant — Hazard mitigation options — Regulatory framework and licensing process for integration of the nuclear power plant with the hydrogen plant — Human resource development (e.g. training)
Codes and standards	— Design codes for a nuclear coupled hydrogen plant (system and components) — Quality specifications for hydrogen — Codes for material selection in hydrogen service — Hydrogen safety codes and availability of associated legislation — Special materials requirements for the hydrogen plant
Environmental aspects	— Project site selection criteria — Emissions and releases from the hydrogen plant (e.g. chemical effluents, by-products, waste heat) — Environmental impacts of emissions and releases — Fresh water withdrawal requirements
Business aspects	— Business models (e.g. hydrogen purchase agreements, by-product oxygen sales) — Identification of the offtake industries — Methods of cost–benefit calculations — Funding and financing options — Subsidies and incentives applicable to the project — Business risk identification and management — Supply chains for materials and components
Policy aspects	— National energy mix and share of renewable and nuclear sources — National hydrogen strategy — Determination of national electricity to hydrogen requirements ratio — Nuclear hydrogen purchase and/or procurement obligations for specific sectors — Project activity timelines — Linkage to programmes for low carbon infrastructure development

7.4. DEPLOYMENT INDICATORS FOR NUCLEAR HYDROGEN PRODUCTION PROJECTS

7.4.1. Deployment indicators

Deployment indicators refer to those factors (qualitative and/or quantitative) that may be used to create a decision matrix for a technological decision [225], which in this case refers to the decision to deploy one or more nuclear hydrogen production projects. An illustrative, non-exhaustive set of such indicators for a country interested in deploying nuclear hydrogen projects (and that is already nuclear power equipped or has made the decision to include nuclear power in its domestic energy mix) is indicated in Table 24, based on considerations identified in Section 7.3 of this publication. The indicators are classified under six broad headings pertaining to the different decision areas in a large scale energy transition. Brief insights into their relevance and significance to the development of a nuclear hydrogen roadmap are provided subsequently. A thorough understanding of these indicators and quantitative determination of the values of these metrics wherever applicable, along with incorporation of the relevant aspects from the IAEA Milestones approach [206], are a crucial part of deciding whether a country wishes to implement these nuclear hydrogen projects and then developing the roadmap for actual commercial scale deployment.

TABLE 24. DEPLOYMENT INDICATORS FOR COUNTRIES PLANNING NUCLEAR HYDROGEN PROGRAMMES

1. Energy use	2. Industries and sectors	3. GHG emissions	4. Hydrogen technologies	5. Economics and finance	6. Regulatory framework
Per capita energy consumption	Industry types and sectors in the economy	Total GHG emissions of the country	Indigenous or imported technologies; extent of localization possible	Levelized cost (production, storage and distribution), cost–benefit ratios, by-product value, system value creation	Nuclear regulation system
Forms of consumption and extent of nuclear power available	Energy forms used; alternatives available	Committed emissions reduction / nationally determined contributions	State of supply chain of systems and components related to hydrogen and its resilience	Sources and allocation of finances to nuclear hydrogen projects; potential business/ revenue models	Availability of licensing scheme for nuclear hydrogen projects
Clean energy options for various sectors and policies (national/ regional)	Potential industrial/ sectoral hydrogen users and near and long term hydrogen demand; hubs/valleys	Sectoral emissions, emissions intensity and methods of reduction available; potential contribution from hydrogen to decarbonization	Options for dispatching hydrogen to connect the point of generation and the point of use	Green finance availability/ green taxonomy considerations	National or international codes and standards for hydrogen service, safety

TABLE 24. DEPLOYMENT INDICATORS FOR COUNTRIES PLANNING NUCLEAR HYDROGEN PROGRAMMES (cont.)

1. Energy use	2. Industries and sectors	3. GHG emissions	4. Hydrogen technologies	5. Economics and finance	6. Regulatory framework
State of energy security of country	Projected industrial growth rates; readiness for switching to hydrogen	Carbon pricing; availability of carbon dioxide capture/ removal technologies	Alternative sources of hydrogen available	Extent of involvement of the private sector in nuclear industry	Certification schemes, guarantees of origin and hydrogen purchase obligations

7.4.2. Significance of the deployment indicators for nuclear hydrogen projects

The energy resource and utilization related deployment indicators in column 1 of Table 24 enable a country to quantify its current and projected energy demands (e.g. through metrics such as primary/final energy consumption per capita, extent of electricity in final consumption, emissions intensity associated with energy consumption and hence economic activities) as a function of the growth and development of its economy; the options available for it to meet current and projected energy demands; and the share of low carbon energy forms available to it for energy transition. This sets the context for adding nuclear power to the country's energy mix.

The indicators in column 2 of Table 24 pertain to energy use patterns from different sectors of the economy as well as the low carbon options available to them. They also identify the sectoral needs for low carbon hydrogen as one of the decarbonization options and the possibility of locating hydrogen production facilities in the vicinity of potential demand centres or consumption sites.

The indicators in column 3 of Table 24 quantify the carbon (and other GHG) emissions intensity of the various activities or sectors contributing to the country's economy, committed reduction in these emissions (such as nationally determined contributions submitted by the government to the United Nations Framework Convention on Climate Change) and the timeframes to act on these commitments; demand from sectors for which hydrogen would be a relevant option for emissions reduction; and mechanisms to finance these reductions.

The indicators in column 4 of Table 24 identify the state of readiness of the country to produce hydrogen for, and distribute the hydrogen to, the relevant sectors; the energy forms (including nuclear power where available) likely to be used for hydrogen production; hydrogen technology availability and extent of localization (both indigenous and imported, including both supply side and demand side technologies); and the supply chains for materials and components relevant to hydrogen technologies.

The indicators in column 5 of Table 24 are the economic and financial metrics useful for hydrogen project evaluation. They include levelized or life cycle costs of production, storage and distribution of hydrogen using different energy forms and feedstock materials; the relative economic and environmental costs and benefits; fiscal incentives (e.g. revoking fossil fuel subsidies, making available concessional finance available under climate finance schemes, inviting foreign investors) that can support hydrogen production; and the costs of alternative GHG abatement technologies that a particular industry might make use of in place of hydrogen. These indicators are meant to support holistic comparisons of the alternatives and provide useful insights for project deployment decision making.

The indicators in column 6 of Table 24 describe the regulatory preparedness to support the growth of a hydrogen economy in general and nuclear hydrogen production in particular. This includes provisions available for certification of low carbon hydrogen, rules and regulations pertaining to safety, quality standards for hydrogen and the availability of frameworks to guide nuclear hydrogen project deployment. The availability of a strong regulatory framework helps to send clear signals about a country's policies and commitment towards hydrogen energy and facilitates financial de-risking of the projects (or at least the first of its kind deployment initiatives).

7.4.3. Opportunities, challenges and questions

Table 25 presents a list of opportunities, challenges and questions to be considered by the key organizations and actors involved in a nuclear hydrogen project. However, this list is not exhaustive.

TABLE 25. OPPORTUNITIES, CHALLENGES AND QUESTIONS TO BE CONSIDERED BY THE KEY ORGANIZATIONS AND ACTORS INVOLVED IN A NUCLEAR HYDROGEN PROJECT

Key organization/ actor	Opportunities	Challenges and questions
NPP operators/ owners	— Operating plants: Hydrogen collocation could be a justification to extend a nuclear power plant's lifetime or renew its licence, as the hydrogen production can benefit from low cost and low carbon electricity/heat from the nuclear reactor and provide an additional revenue stream for the nuclear power plant. — Financial aspect: Diversification of revenues from hydrogen production, as well as storage (where applicable). — New plants: Hydrogen production could be integrated into the new designs from an early stage, with the possibility of optimization for hydrogen production.	— Operating plant: How to assess risks with respect to remaining lifetime and investment trade-off? — Licensing: Is the non-nuclear licensing or current nuclear licensing sufficient for the nuclear hydrogen licensing? — Security considerations: Hydrogen plant requiring separate physical protection; new staff who may not work full time at the plant and who might need separate training and operator certification, etc.
Nuclear technology and hydrogen technology developers	— Funding: Favourable government policies could significantly reduce the burden associated with necessary funding, facilitating the nuclear technology developers to move ahead with their design and obtain licensing. — Time frames: Possibility of engaging regulators in reviewing the nuclear hydrogen production plant project at an early stage.	— Time frames: Time is required by the regulator to review and approve the adjustments to the project (the coupling of the hydrogen production plant with the nuclear power plant). — Hydrogen plant technology suppliers need to be prepared to understand the modifications that might be needed in design codes, fabrication standards, testing and quality assurance requirements for the components of the hydrogen plant when it has to be coupled with a nuclear power plant.

TABLE 25. OPPORTUNITIES, CHALLENGES AND QUESTIONS TO BE CONSIDERED BY THE KEY ORGANIZATIONS AND ACTORS INVOLVED IN A NUCLEAR HYDROGEN PROJECT (cont.)

Key organization/ actor	Opportunities	Challenges and questions
Regulatory bodies	— International collaboration: Experience sharing and regulatory harmonization through international collaboration. — Development of necessary regulations: Development of a phase wise regulatory mechanism with a graded approach and early engagement with the regulator could save time during detailed review periods.	— New technology: The regulation of a nuclear reactor coupled with a hydrogen production facility has implications on the nuclear regulator when the hydrogen production facility is collocated with the nuclear power plant, as the hydrogen production facility can impact the safety of the nuclear power plant; if not collocated with the nuclear power plant, the hydrogen production facility can be treated as a regular energy consumer: • How will reviews be completed? • How to harmonize necessary regulations as approaches to codes and standards for both nuclear and hydrogen fields vary in different countries?

TABLE 25. OPPORTUNITIES, CHALLENGES AND QUESTIONS TO BE CONSIDERED BY THE KEY ORGANIZATIONS AND ACTORS INVOLVED IN A NUCLEAR HYDROGEN PROJECT (cont.)

Key organization/ actor	Opportunities	Challenges and questions
Government	— Policy support: Financial policies and clean energy standards are prerequisites to large scale, broad sector decarbonization using nuclear power. Technology neutral policy making will be needed and all forms of low carbon hydrogen production processes have to be given similar incentives. — Net zero goals: Nuclear hydrogen technologies can contribute to achieving net zero goals. — Hybrid systems: Foster understanding of the benefits of renewables and nuclear partnership in an integrated grid as well as in the context of hydrogen production.	— Public acceptance: Acceptance of a ramp-up of the hydrogen market has to be sought. — Additional policies: For example, at large scale, water policy consideration may be necessary. — Tax policy: Introduction of a carbon tax, removal of subsidies to fossil fuel based hydrogen production (or vice versa, support for hydrogen producers) and the definition of tax parameters that may differ for the different hydrogen production options.
Hydrogen producers and users	— New markets: Nuclear, hydrogen production can enable the project developers to diversify into a new market but backed up by a mature technology.	— Siting options: New reactor designs may add flexibility to overcome the siting challenges and speed up the siting process. A location closer to large industrial users may be possible, thus minimizing the need for a transportation and distribution infrastructure. — Safety: How to regulate hydrogen safety (explosion) with large volumes of hydrogen production and its storage near the production site.

TABLE 25. OPPORTUNITIES, CHALLENGES AND QUESTIONS TO BE CONSIDERED BY THE KEY ORGANIZATIONS AND ACTORS INVOLVED IN A NUCLEAR HYDROGEN PROJECT (cont.)

Key organization/ actor	Opportunities	Challenges and questions
Renewable sector	— Enhancing the technology: • Partnering renewables and nuclear energy will address variability and ensure grid/ system stability and high annual capacity utilization of the hydrogen plants. • Energy generation hubs can offset risk, improving reliability for hydrogen customers. — Funding: Hybrid energy systems that include nuclear energy for hydrogen production can benefit from attractive funding partnerships. — Net zero goals: Partnerships with the renewables sector may provide benefit from attractive government policies.	— Competition: Nuclear power programmes are frequently perceived to be a competitor for subsidies and political support, etc.
Public	— Partnership nuclear–renewable energies: Foster an understanding of the benefits of renewables and nuclear partnership in an integrated decarbonized grid as well as in the context of hydrogen production for industrial decarbonization.	— Lack of general understanding of the need for baseload electricity supply, and roles/ benefits of nuclear energy in a diverse, integrated and reliable low carbon energy system.

7.5. LINKAGE WITH OTHER DECARBONIZATION ROADMAPS

A nuclear hydrogen roadmap cannot be developed in isolation, as it needs to have strong interlinkages with short and long term national energy policies, decarbonization/net zero strategies and the sustainable development agenda of the country concerned, as shown by the deployment indicators discussed in Section 7.4. It also needs to be directly related to the overall national hydrogen strategy of the country, which has to consider nuclear hydrogen as one of the low carbon hydrogen options available to it, alongside alternatives such as renewable hydrogen or hydrogen from biomass with carbon capture.

A publication of a consortium of academia, industry and international organizations — the Nuclear Hydrogen Initiative — describes that currently more than 50 countries have a declared national hydrogen strategy, and about 15% of them have explicitly stated plans for using their nuclear power assets (existing or upcoming) to produce hydrogen [226]. Technologies of choice for the first of its kind nuclear hydrogen production programmes among Member States include water and steam electrolysis for coupling with water cooled reactors for the near term (e.g. Canada, UK, USA) and thermochemical processes for the intermediate to long term, based on the maturity of advanced high temperature reactor technologies (e.g. Japan, UK).

The combination of hydrogen production technology and nuclear reactor technology is an important aspect that has to be assessed critically by a country considering nuclear hydrogen project deployment. Particularly for a first of its kind project of this nature, the availability of technically and commercially mature alternatives will set the timeline by which project deployment may be expected to be completed.

7.6. NUCLEAR HYDROGEN ROADMAP PROJECT

Table 26 presents the roadmap for a nuclear hydrogen project, building on the previous sections of this publication and addressing the following dimensions:

- The national policy framework for nuclear hydrogen programmes (see Sections 2, 3 and 7);
- Nuclear technology selection (see Section 4);
- Hydrogen technology selection (see Section 4);
- Coupling schemes (see Sections 3 and 4);
- Safety aspects (see Section 5);
- Regulations, codes and standards (see Section 5);
- Market creation and other commercial aspects (see Section 3);

— Stakeholder engagement (see Section 6).

The major considerations for the workstreams are included, as well as the specific tasks to be addressed under the different phases of the project:

— Phase 1: Planning of policy and technology;
— Phase 2: Development of technology and infrastructure;
— Phase 3: Demonstration of technology and industry uses;
— Phase 4: Sustaining a long term nuclear hydrogen programme.

The major project milestones based on the activities identified in the roadmap are the following:

— Identification of the nuclear power plant site for the nuclear hydrogen project;
— Identification of the hydrogen technology and vendor selection;
— Decision on the final investment for coupling with the nuclear reactor;
— Identification of the hydrogen end users;
— Finalization of the reactor to hydrogen plant coupling scheme and the hydrogen plant to hydrogen user coupling scheme;
— Completion of stakeholder engagement and environmental impact assessment studies;
— Licensing of the design and receipt of regulatory clearance for plant construction;
— Construction and commissioning of the plant;
— Production and commercial dispatch of the product to the end user.

Text cont. on p. 145.

TABLE 26. ROADMAP FOR NUCLEAR HYDROGEN PRODUCTION

Broad workstream	Major considerations under the workstream	Phase 1: Planning of policy and technology	Phase 2: Development of technology and infrastructure	Phase 3: Demonstration of technology and industry uses	Phase 4: Sustaining a long term nuclear hydrogen programme
1. National policy framework for nuclear hydrogen programmes	Decision making and commitment to the use of nuclear power for the hydrogen programme as part of a decarbonization strategy.	Development of national hydrogen strategies as part of decarbonization and national energy security policies.	Development and implementation of any incentives, concessions for nuclear hydrogen projects (e.g. tax incentives, concessional financing schemes).	Implementation of further incentives, concessions for nuclear hydrogen projects and creation of additional policy measures to create long term markets for hydrogen usage.	Providing support for future nuclear hydrogen projects, capacity building and resource allocation.
	Identification of further R&D needs specific to nuclear hydrogen production.	Identification of: — Short and long term low carbon hydrogen needs of the country/ region; — Potential users of hydrogen; — Diversified sources of hydrogen.			

TABLE 26. ROADMAP FOR NUCLEAR HYDROGEN PRODUCTION (cont.)

Broad workstream	Major considerations under the workstream	Phase 1: Planning of policy and technology	Phase 2: Development of technology and infrastructure	Phase 3: Demonstration of technology and industry uses	Phase 4: Sustaining a long term nuclear hydrogen programme
	Policy for programme expansion.	Providing long term support towards developing/expanding nuclear power programmes as part of decarbonization goals.			
	Timelines for programme initiation.	Recognition of and commitment to nuclear hydrogen production as a route towards diversification of hydrogen sources available to the country.			
2. Nuclear technology selection	Use of existing or mature reactor technologies (e.g. LWRs, PHWRs) by retrofitting projects.	Taking decision regarding the adoption of specific nuclear technology.			Marketing and production of hydrogen from first of its kind projects.

TABLE 26. ROADMAP FOR NUCLEAR HYDROGEN PRODUCTION (cont.)

Broad workstream	Major considerations under the workstream	Phase 1: Planning of policy and technology	Phase 2: Development of technology and infrastructure	Phase 3: Demonstration of technology and industry uses	Phase 4: Sustaining a long term nuclear hydrogen programme
	Use of existing or mature reactor technologies (e.g. LWRs, PHWRs) through new build projects.	Taking decision regarding the nuclear technology to be used to support hydrogen production.	Selection of the project deployment site.	Establishment of the supply chain for systems and components of hydrogen plants.	
	Use of advanced technologies (e.g. SMRs, advanced reactors).	Taking decision regarding feedstock for nuclear hydrogen production.	Development of the conceptual design of the facility.	Establishment of the supply chain, business model, detailed design, licensing, construction, operation of the demonstration plant.	Developing further commercial projects with or without technology upgrades, sustained marketing.

TABLE 26. ROADMAP FOR NUCLEAR HYDROGEN PRODUCTION (cont.)

Broad workstream	Major considerations under the workstream	Phase 1: Planning of policy and technology	Phase 2: Development of technology and infrastructure	Phase 3: Demonstration of technology and industry uses	Phase 4: Sustaining a long term nuclear hydrogen programme
3. Hydrogen technology selection	Mature technology to be coupled with nuclear reactors: Reforming processes (e.g. methane reforming, LTE).	Selection of: — Hydrogen production technology based on selected nuclear reactor technology; — Storage technology (long/short term); — Distribution technology to specific end user.	Finalization of hydrogen technology, end user of hydrogen, scale of operation, scheme for product dispatch to user.	Fabrication of plant components at the vendor's site.	Arrangement of the long term supply between the end user and the NPP owner/operator.
	Advanced technologies to be coupled with nuclear reactors: high temperature SOEC, other cycles.	Identification of the technology supplier/vendor.	Development of sustainable supply chain of hydrogen production, storage and distribution technologies.	On-site integration of HPP with the NPP.	

TABLE 26. ROADMAP FOR NUCLEAR HYDROGEN PRODUCTION (cont.)

Broad workstream	Major considerations under the workstream	Phase 1: Planning of policy and technology	Phase 2: Development of technology and infrastructure	Phase 3: Demonstration of technology and industry uses	Phase 4: Sustaining a long term nuclear hydrogen programme
	Advanced technologies: thermochemical, hybrid cycles.	Identification of freshwater production technology for providing feed to the hydrogen plant; identification of heat and electricity supply schemes to the hydrogen plant.	Taking final investment decision, approval for project funding and sanction of funds.	Gather operator feedback and development of standard operating procedure	
	Advanced technology: methane pyrolysis	Site selection for nuclear hydrogen project (i.e. first of its kind and nth of its kind).		Ensuring the necessary technology upgrades during replacements.	
4. Coupling schemes	Determination of the coupling scheme based on the reactor and the hydrogen technology selected.	Planning for electrical coupling only (e.g. for LTE)	Identification of alternative/back up H_2 sources for end user industries in case of NPP downtime.	Establishing of emergency shutdown procedures for hydrogen plant according to the nature of the coupling.	

TABLE 26. ROADMAP FOR NUCLEAR HYDROGEN PRODUCTION (cont.)

Broad workstream	Major considerations under the workstream	Phase 1: Planning of policy and technology	Phase 2: Development of technology and infrastructure	Phase 3: Demonstration of technology and industry uses	Phase 4: Sustaining a long term nuclear hydrogen programme
	Determination of the coupling scheme between the hydrogen producer and the user.	Planning for electrical and thermal coupling (e.g. for thermochemical cycles and HTE).	Finalization of the coupling scheme between the NPP and the hydrogen user (e.g. need for intermediate hydrogen storage and its sizing, layout of pipelines).	Establishing of emergency evacuation procedures in case of off-site emergencies.	
	Considerations on the adoption of collocation or coupling for the project.	Planning for loosely coupled or tightly coupled plant and preliminary facility layout.			
5. Safety aspects	Safety considerations arising from the hydrogen plant and potential impact on reactor operation.	Identification of additional safety hazards/external events for a collocated hydrogen plant (e.g. hydrogen fire, leakage, toxic gas release).	Performing detailed safety analysis (deterministic/ probabilistic) based on regulatory requirements.	Ensuring safety in construction and other on-site activities at the NPP.	Ensuring safety in major replacement activities.

TABLE 26. ROADMAP FOR NUCLEAR HYDROGEN PRODUCTION (cont.)

Broad workstream	Major considerations under the workstream	Phase 1: Planning of policy and technology	Phase 2: Development of technology and infrastructure	Phase 3: Demonstration of technology and industry uses	Phase 4: Sustaining a long term nuclear hydrogen programme
	Impact of reactor operational conditions (e.g. transients) on the hydrogen plant.	Conducting preliminary accident analysis based on identified hazards for planning facility layout, mitigation measures, evacuation plans.	Development of hazard prevention plans and mitigation measures.	Conducting plant start-up and cold commissioning.	Ensuring safety in plant decommissioning activities.
		Engagement with nuclear and other regulators to develop an outline of a licensing scheme for nuclear hydrogen projects.		Establishing regular operation philosophy and safety philosophy.	
6. Regulations, codes and standards	Hydrogen safety codes and standards.	Availability/adoption of codes and standards for hydrogen plants (e.g. purity, materials selection, equipment design, safety related, hydrogen transportation related).	Conducting licensing activities (e.g. site clearance and design certification).	Issuance of construction permits.	Developing in-service inspection schemes (e.g. hydrogen embrittlement of components); periodic regulatory review and relicensing procedure for the hydrogen plant.

TABLE 26. ROADMAP FOR NUCLEAR HYDROGEN PRODUCTION (cont.)

Broad workstream	Major considerations under the workstream	Phase 1: Planning of policy and technology	Phase 2: Development of technology and infrastructure	Phase 3: Demonstration of technology and industry uses	Phase 4: Sustaining a long term nuclear hydrogen programme
	Hydrogen plant construction codes and standards.	Availability/adoption of certification standards for low carbon hydrogen (e.g. based on life cycle carbon emissions intensity of the value chain).	Conducting environmental impact assessment of the nuclear hydrogen plant.	Conducting periodic regulatory inspections — schedule and protocols.	
	Standardized licensing process for hydrogen plants.		Definition of on-site and off-site emergencies and development of an action plan to manage them.	Finalization and implementation of procedures for: — In-service inspection; — Major revamp/ replacement plans; — Decommissioning plans.	

TABLE 26. ROADMAP FOR NUCLEAR HYDROGEN PRODUCTION (cont.)

Broad workstream	Major considerations under the workstream	Phase 1: Planning of policy and technology	Phase 2: Development of technology and infrastructure	Phase 3: Demonstration of technology and industry uses	Phase 4: Sustaining a long term nuclear hydrogen programme
7. Market creation and other commercial aspects	Decision on nuclear hydrogen production for industrial consumers (commodity markets) and/or long term energy storage and/or grid balancing services (energy markets/ancillary services market).	Establishing nuclear hydrogen quotas for large industrial consumers in hard to abate sectors. Site selection and industrial user identification for the creation of nuclear hydrogen user hubs.	Conducting detailed techno-commercial studies on life cycle basis.	Issuance of the contract to the hydrogen technology vendor.	Synergies between nuclear and other low carbon hydrogen production.
	Recognizing clean energy characteristics of nuclear hydrogen in funding sources and final investment decisions.	Identification of sources of funding for project (e.g. national budgets or special instruments such as climate finance, transition finance).	Conducting tendering/ bidding process and negotiations; vendor development for localization of technology.	Developing of performance linked payment policies to the hydrogen technology vendor.	Integration with other market based mechanisms developed for hydrogen (e.g. auctioning, merit order dispatch).
	Supply chains for reactor and hydrogen plant technologies.	Identification of revenue streams and structuring long term contracts with end users of nuclear hydrogen.	Finalization of contract terms with the hydrogen technology vendor.	Achieving first product dispatch to the hydrogen user.	

TABLE 26. ROADMAP FOR NUCLEAR HYDROGEN PRODUCTION (cont.)

Broad workstream	Major considerations under the workstream	Phase 1: Planning of policy and technology	Phase 2: Development of technology and infrastructure	Phase 3: Demonstration of technology and industry uses	Phase 4: Sustaining a long term nuclear hydrogen programme
	Defining the nuclear hydrogen ecosystem and development of nuclear hydrogen value chains.	Establishing additional measures to de-risk the project.	Finalization of contract terms and off take agreements (i.e. hydrogen purchase agreement) including tariff structure with hydrogen end user.		
	Certification schemes for low carbon hydrogen production.	Conduct of initial techno-economic studies and development of the business model.			
8. Stakeholder engagement	Stakeholder mapping.	Creation of the nuclear hydrogen project team. Development of the stakeholder map.	Conduct of public consultation/hearings as part of the stakeholder engagement and regulatory clearance.	Training for hydrogen plant operation, on-site and off-site emergency management.	Establishing long term employment opportunities in the continuing programme.

TABLE 26. ROADMAP FOR NUCLEAR HYDROGEN PRODUCTION (cont.)

Broad workstream	Major considerations under the workstream	Phase 1: Planning of policy and technology	Phase 2: Development of technology and infrastructure	Phase 3: Demonstration of technology and industry uses	Phase 4: Sustaining a long term nuclear hydrogen programme
	Communication strategy and key discussion points.	Development of the communication policy with stakeholders depending on interest and influence levels.			
		Initiation of communication with stakeholders on relevant aspects such as technology, economics, safety, socioeconomic benefits, environmental aspects.			

Note: HPP — hydrogen production plant; LWR — light water reactor; NPP — nuclear power plant; PHWR — pressurized heavy water reactor.

7.7. CONSIDERATIONS AND CONCLUDING REMARKS

A roadmap for nuclear hydrogen production processes might be designed as a decision making support tool for the planning agencies and decision makers of the country. Therefore, it needs to provide a short and long term focus on several cross-cutting issues. It might attempt to be generic and lay out all the various possible courses of action under different workstreams, so that individual project developers can perform assessments, derive useful insights and finally select the best alternatives keeping in mind their specific considerations. The generic roadmap presented in Table 26 lists several relevant issues for consideration by the policy makers and relevant authorities and is therefore intended to be used as a practical guidance tool for interested countries, technology partners and potential investors.

The salient considerations for the decision making bodies relating to the practical realization of nuclear hydrogen projects may be presented as follows:

- The decision to deploy a nuclear hydrogen project coupled with an existing or upcoming nuclear power reactor has to be taken while keeping the broad energy needs, decarbonization objectives and climate change mitigation and adaptation goals of the concerned country in mind. Many such factors are indicated in Table 24.
- Once the final decision on project deployment is reached, detailed consideration of the factors in Table 23 is required. Specific activities to be taken up in each phase (1–4) of project deployment are detailed in Table 26. Major project milestones to track progress are provided in Section 7.6.
- The first of its kind nuclear hydrogen project may be deployed as a pilot initiative to understand the technical and regulatory aspects of such programmes, including development of a graded licensing approach for the projects. Sections 4 and 5 provide guidance on these aspects.
- Subsequent expansion of these projects will have to be based on a strong business case that includes, among others, a clearly identified commercial value proposition, a well-defined market and sustained 'offtakers' for the produced hydrogen, sources of funding and financing, sharing of business risks, stable business model, reliable supply chain for systems and components, and clear benefits including decarbonization, industrial growth and availability of jobs. Section 3 provides detailed information on the key commercialization aspects.
- Stakeholders need to be mapped very early in the project development stage (for an indicative list, see Fig. 27) and channels of communication have to be kept open throughout. Indicative questions to pursue as part of stakeholder engagement are provided in Tables 20 and 21 in Section 6.

REFERENCES

[1] GE, L., et al., A review of hydrogen generation, storage, and applications in power system, J. Energy Storage **75** (2024), https://doi.org/10.1016/j.est.2023.109307

[2] HYDROGEN COUNCIL, Hydrogen Insights 2022: An Updated Perspective on Hydrogen Market Development and Actions Required to Unlock Hydrogen at Scale (2022), https://hydrogencouncil.com/wp-content/uploads/2022/09/Hydrogen-Insights-2022-2.pdf

[3] HYDROGEN COUNCIL, Hydrogen for Net-Zero: A Critical Cost-competitive Energy Vector (2021), https://hydrogencouncil.com/wp-content/uploads/2021/11/Hydrogen-for-Net-Zero.pdf

[4] INTERNATIONAL ATOMIC ENERGY AGENCY, Hydrogen Production with Operating Nuclear Power Plants: Business Case, IAEA, Vienna (2023).

[5] EUROPEAN COMMISSION, Causes of Climate Change (2025), https://climate.ec.europa.eu/climate-change/causes-climate-change_en

[6] REKKER, S.A.C., O'BRIEN, K.R., HUMPHREY, J.E., PASCALE, A.C., Comparing extraction rates of fossil fuel producers against global climate goals, Nat. Clim. Change **8** 6 (2018) 489–492, https://doi.org/10.1038/s41558-018-0158-1

[7] INTERGOVERNMENTAL PANEL ON CLIMATE CHANGE, Climate Change 2021 — The Physical Science Basis: Working Group I Contribution to the Sixth Assessment Report of the Intergovernmental Panel on Climate Change, Cambridge University Press, Cambridge (2023), https://doi.org/10.1017/9781009157896

[8] INTERGOVERNMENTAL PANEL ON CLIMATE CHANGE, Climate Change 2022 — Mitigation of Climate Change: Working Group III Contribution to the Sixth Assessment Report of the Intergovernmental Panel on Climate Change, Cambridge University Press, Cambridge (2023), https://doi.org/10.1017/9781009157926

[9] GE, M., FRIEDRICH, J., VIGNA, L., Where Do Emissions Come From? 4 Charts Explain Greenhouse Gas Emissions by Sector, World Resources Institute (2024), https://www.wri.org/insights/4-charts-explain-greenhouse-gas-emissions-countries-and-sectors

[10] INTERNATIONAL ENERGY AGENCY, Net Zero by 2050 — A Roadmap for the Global Energy Sector, IEA, Paris (2021), https://www.iea.org/reports/net-zero-by-2050

[11] INTERNATIONAL RENEWABLE ENERGY AGENCY, World Energy Transitions Outlook 2022, IRENA, Abu Dhabi (2022).

[12] INTERNATIONAL ATOMIC ENERGY AGENCY, Climate Change and Nuclear Power 2020, IAEA, Vienna (2020).

[13] BROUWER, K.M., BERGKAMP, L., Road to EU Climate Neutrality by 2020: Spatial Requirements of Wind/Solar and Nuclear Energy and Their Respective Costs, ECR Group and Renew Europe, Brussels (2021),
https://roadtoclimateneutrality.eu/Energy_Study_Full.pdf

[14] INTERNATIONAL ENERGY AGENCY, The Role of Critical Minerals in Clean Energy Transitions, IEA, Paris (2021),
https://www.iea.org/reports/the-role-of-critical-minerals-in-clean-energy-transitions

[15] INTERNATIONAL ENERGY AGENCY, Electricity 2024: Analysis and Forecast to 2026, IEA, Paris (2024),
https://www.iea.org/reports/electricity-2024

[16] COMMITTEE ON CLIMATE CHANGE, Hydrogen in a Low-Carbon Economy, London (2018),
https://www.theccc.org.uk/wp-content/uploads/2018/11/Hydrogen-in-a-low-carbon-economy.pdf

[17] INGERSOLL, E., GOGAN, K., Missing Link to a Livable Climate: How Hydrogen-Enabled Synthetic Fuels Can Help Deliver the Paris Goals, LucidCatalyst (2020),
https://www.lucidcatalyst.com/_files/ugd/2fed7a_f532181b09b2455f9f5eecfc97cc58ea.pdf

[18] FRAZER NASH CONSULTANCY, Fugitive Hydrogen Emissions in a Future Hydrogen Economy, FNC 012865-53172R, Issue 1 (2022),
https://assets.publishing.service.gov.uk/media/624ec79cd3bf7f600d4055d1/fugitive-hydrogen-emissions-future-hydrogen-economy.pdf

[19] LIEBREICH, M., The Clean Hydrogen Ladder, V4.1, Liebreich Associates (2021),
https://www.liebreich.com/the-clean-hydrogen-ladder-now-updated-to-v4-1/

[20] MARCHESE, M., GANDIGLIO, M., LANZINI, A., A circular approach for making Fischer–Tropsch e-fuels and e-chemicals from biogas plants in Europe, Front. Energy Res. **9** (2021),
https://doi.org/10.3389/fenrg.2021.773717

[21] UECKERDT, F., et al., Potential and risks of hydrogen-based e-fuels in climate change mitigation, Nat. Clim. Change **11** 5 (2021) 384–393,
https://doi.org/10.1038/s41558-021-01032-7

[22] RAM, V., SALKUTI, S.R., An overview of major synthetic fuels, Energies **16** 6 (2023) 2834,
https://doi.org/10.3390/en16062834

[23] GUERRA, O.J., et al., The value of seasonal energy storage technologies for the integration of wind and solar power, Energy Environ. Sci. **13** 7 (2020) 1909–1922,
https://doi.org/10.1039/D0EE00771D

[24] KOMOROWSKA, A., OLCZAK, P., HANC, E., KAMIŃSKI, J., An analysis of the competitiveness of hydrogen storage and Li-ion batteries based on price arbitrage in the day-ahead market, Int. J. Hydrogen Energy **47** 66 (2022) 28556–28572,
https://doi.org/10.1016/j.ijhydene.2022.06.160

[25] MAYRHOFER, M., KOLLER, M., SEEMANN, P., PRIELER, R., HOCHENAUER, C., Assessment of natural gas/hydrogen blends as an alternative fuel for industrial heat treatment furnaces, Int. J. Hydrogen Energy **46** 41 (2021) 21672–21686, https://doi.org/10.1016/j.ijhydene.2021.03.228

[26] WANG, R.R., ZHAO, Y.Q., BABICH, A., SENK, D., FAN, X.Y., Hydrogen direct reduction (H-DR) in steel industry — An overview of challenges and opportunities, J. Cleaner Prod. **329** (2021), https://doi.org/10.1016/j.jclepro.2021.129797

[27] VOGL, V., ÅHMAN, M., NILSSON, L.J., Assessment of hydrogen direct reduction for fossil-free steelmaking, J. Cleaner Prod. **203** (2018) 736–745, https://doi.org/10.1016/j.jclepro.2018.08.279

[28] CHIESA , P., LOZZA, G., MAZZOCCHI, L., Using hydrogen as gas turbine fuel, J. Eng. Gas Turbines Power **127** 1 (2005) 73–80, https://doi.org/10.1115/1.1787513

[29] ATILHAN, S., et al., Green hydrogen as an alternative fuel for the shipping industry, Curr. Opin. Chem. Eng. P**31** (2021), https://doi.org/10.1016/j.coche.2020.100668

[30] EUROPEAN COMMISSION, Development of a Large Fuel Cell Stack for Maritime Applications (2022), https://cordis.europa.eu/programme/id/HORIZON_HORIZON-JTI-CLEANH2-2023-03-02

[31] ELKAFAS, A.G., RIVAROLO, M., GADDUCCI, E., MAGISTRI, L., MASSARDO, A.F., Fuel cell systems for maritime: A review of research development, commercial products, applications, and perspectives, Processes **11** 1 (2023), https://doi.org/10.3390/pr11010097

[32] MELAINA, M.W., ANTONIA, A., PENEV, M., Blending Hydrogen into Natural Gas Pipeline Networks: A Review of Key Issues, Rep. NREL/TP-5600-51995, National Renewable Energy Laboratory, Colorado (2013), https://doi.org/10.2172/1068610

[33] EUROPEAN HYDROGEN BACKBONE, Analysing Future Demand, Supply, and Transport of Hydrogen, Utrecht (2021), https://ehb.eu/files/downloads/EHB-Analysing-the-future-demand-supply-and-transport-of-hydrogen-June-2021-v3.pdf

[34] INTERNATIONAL ENERGY AGENCY, Global Hydrogen Demand by Sector in the Sustainable Development Scenario, 2019–2070, IEA, Paris (2020), https://www.iea.org/data-and-statistics/charts/global-hydrogen-demand-by-sector-in-the-sustainable-development-scenario-2019-2070

[35] INTERNATIONAL RENEWABLE ENERGY AGENCY, Global Hydrogen Trade to Meet the 1.5°C Climate Goal: Part III — Green Hydrogen Cost and Potential, IRENA, Abu Dhabi (2022), https://www.irena.org/publications/2022/May/Global-hydrogen-trade-Cost

[36] TONELLI, D., et al., Global land and water limits to electrolytic hydrogen production using wind and solar resources, Nat. Commun. **14** 1 (2023) 5532, https://doi.org/10.1016/j.coche.2020.100668

[37] UNITED NATIONS CLIMATE CHANGE, Key Aspects of the Paris Agreement, https://unfccc.int/most-requested/key-aspects-of-the-paris-agreement

[38] INTERNATIONAL RENEWABLE ENERGY AGENCY, World Energy Transitions Outlook 2023: 1.5°C Pathway, IRENA, Abu Dhabi (2023), https://www.irena.org/-/media/Files/IRENA/Agency/Publication/2023/Mar/IRENA_WETO_Preview_2023.pdf

[39] DNV, Hydrogen Forecast to 2050: Energy Transition Outlook 2022, DNV AS, Høvik (2022), https://www.dnv.com/focus-areas/hydrogen/forecast-to-2050/

[40] NATERER, G.F., JABER, O., DINCER, I., "Environmental impact comparison of steam methane reformation and thermochemical processes of hydrogen production" (Proc. 18th WHEC, Essen, 2010) (2010), https://juser.fz-juelich.de/record/135464/files/HP4a_3_Naterer_rev0601.pdf

[41] HOWARTH, R.W., JACOBSON, M.Z., How green is blue hydrogen? Energy Sci. Eng. **9** 10 (2021) 1676–1687, https://doi.org/10.1002/ese3.956

[42] DAGDOUGI, H., SACILE, R., BERSANI, C., OUAMMI, A., Hydrogen Infrastructure for Energy Applications: Production, Storage, Distribution and Safety, Academic Press, London (2018).

[43] MEERMAN, J.C., et al., Techno-economic assessment of CO2 capture at steam methane reforming facilities using commercially available technology, Int. J. Greenhouse Gas Control **9** (2012) 160–171, https://doi.org/10.1016/j.ijggc.2012.02.018

[44] INTERNATIONAL ORGANIZATION FOR STANDARDIZATION, Environmental Management — Life Cycle Assessment — Principles and Framework, ISO 14040:2006, ISO, Geneva (2006).

[45] INTERNATIONAL ORGANIZATION FOR STANDARDIZATION, Environmental Management — Life Cycle Assessment — Requirements and Guidelines, ISO 14044:2006, ISO, Geneva (2006).

[46] WANG, N., AKIMOTO, K., NEMET, G.F., What went wrong? Learning from three decades of carbon capture, utilization and sequestration (CCUS) pilot and demonstration projects, Energy Policy **158** (2021), https://doi.org/10.1016/j.enpol.2021.112546

[47] CHANG, A.C.C., CHANG, H.-F., LIN, F.-J., LIN, K.-H., CHEN, C.-H., Biomass gasification for hydrogen production, Int. J. Hydrogen Energy **36** 21 (2011) 14252–14260, https://doi.org/10.1016/j.ijhydene.2011.05.105

[48] STIEGEL, G.J., RAMEZAN, M., Hydrogen from coal gasification: An economical pathway to a sustainable energy future, Int. J. Coal Geol. **65** 3 (2006) 173–190, https://doi.org/10.1016/j.ijhydene.2011.05.105

[49] SÁNCHEZ-BASTARDO, N., SCHLÖGL, R., RULAND, H., Methane pyrolysis for zero-emission hydrogen production: A potential bridge technology from fossil fuels to a renewable and sustainable hydrogen economy, Ind. Eng. Chem. Res. **60** 32 (2021) 11855–11881,
https://doi.org/10.1021/acs.iecr.1c01679

[50] ZENG, J., et al., Catalytic methane pyrolysis with liquid and vapor phase tellurium, ACS Catal. **10** 15 (2020) 8223–8230,
https://doi.org/10.1021/acscatal.0c00805

[51] INTERNATIONAL ENERGY AGENCY, Global Hydrogen Review 2021, IEA, Paris (2021),
https://iea.blob.core.windows.net/assets/3a2ed84c-9ea0-458c-9421-d166a9510bc0/GlobalHydrogenReview2021.pdf

[52] INTERNATIONAL ENERGY AGENCY, The Future of Hydrogen — Seizing Today's Opportunities, IEA, Paris (2019),
https://www.iea.org/reports/the-future-of-hydrogen

[53] WARWICK, N., et al., Atmospheric Implications of Increased Hydrogen Use, OGL (2022).

[54] INTERNATIONAL RENEWABLE ENERGY AGENCY, Geopolitics of the Energy Transformation: The Hydrogen Factor, IRENA, Abu Dhabi (2022),
https://www.irena.org/Publications/2022/Jan/Geopolitics-of-the-Energy-Transformation-Hydrogen

[55] HANDA, R., Geopolitical Implications of Green Hydrogen Economy, Observer Research Foundation (2021),
https://www.orfonline.org/expert-speak/geopolitical-implications-of-green-hydrogen-economy/

[56] CLEAN AIR TASK FORCE, Hydrogen for Decarbonization: A Realistic Assessment (2023),
https://www.catf.us/resource/hydrogen-decarbonization-realistic-assessment/

[57] GBADEYAN, O.J., MUTHIVHI, J., LINGANISO, L.Z., DEENADAYALU, N., Decoupling economic growth from carbon emissions: A transition toward low-carbon energy systems — A critical review, Clean Technol. (2024) 1076–1113,
https://doi.org/10.3390/cleantechnol6030054

[58] INTERNATIONAL ENERGY AGENCY, Global Hydrogen Review 2022, IEA, Paris (2022),
https://iea.blob.core.windows.net/assets/c5bc75b1-9e4d-460d-9056-6e8e626a11c4/GlobalHydrogenReview2022.pdf

[59] ENERGY TRANSITION COMMISSION, Making the Hydrogen Economy Possible: Accelerating Clean Hydrogen in an Electrified Economy, ETC, London (2021).

[60] HYDROGEN COUNCIL, Hydrogen Decarbonization Pathways, Hydrogen Council, Brussels (2021).

[61] INTERNATIONAL ENERGY AGENCY, Global Hydrogen Review 2023, IEA, Paris (2023),
https://iea.blob.core.windows.net/assets/cb9d5903-0df2-4c6c-afa1-4012f9ed45d2/GlobalHydrogenReview2023.pdf

[62] AHMED, S.F., et al., Sustainable hydrogen production: Technological advancements and economic analysis, Int. J. Hydrogen Energy **47** 88 (2022) 37227–37255, https://doi.org/10.1016/j.ijhydene.2021.12.029
[63] WORLD INTEGRATED TRADE SOLUTION, Hydrogen Exports by Country in 2019 (2019), https://wits.worldbank.org/trade/comtrade/en/country/ALL/year/2019/tradeflow/Exports/partner/WLD/product/280410
[64] KORBERG, A.D., et al., On the feasibility of direct hydrogen utilisation in a fossil-free Europe, Int. J. Hydrogen Energy **48** 8 (2023) 2877–2891, https://doi.org/10.1016/j.ijhydene.2022.10.170
[65] PARGAR, F., KUJALA, J., AALTONEN, K., RUUTU, S., Value creation dynamics in a project alliance, Int. J. Proj. Manag. **37** 5 (2019) 716–730, https://doi.org/10.1016/j.ijhydene.2022.10.170
[66] WOJEWNIK-FILIPKOWSKA, A., DZIADKIEWICZ, A., DRYL, W., DRYL, T., BĘBEN, R., Obstacles and challenges in applying stakeholder analysis to infrastructure projects, J. Prop. Invest. Finance **39** 3 (2021) 199–222, https://doi.org/10.1108/JPIF-03-2019-0037
[67] INTERNATIONAL ATOMIC ENERGY AGENCY, Nuclear Energy for a Net Zero World, IAEA, Vienna (2021), https://www.iaea.org/sites/default/files/21/10/nuclear-energy-for-a-net-zero-world.pdf
[68] DI MADDALONI, F., DAVIS, K., The influence of local community stakeholders in megaprojects: Rethinking their inclusiveness to improve project performance, Int. J. Proj. Manag. **35** 8 (2017) 1537–1556, https://doi.org/10.1016/j.ijproman.2017.08.011
[69] LYONS, M., DURRANT, P., KOCHHAR K., Reaching Zero with Renewables Capturing Carbon, International Renewable Energy Agency, Abu Dhabi (2021), https://www.irena.org/-/media/Irena/Files/Technical-papers/IRENA_Capturing_Carbon_2021.pdf?rev=bf05359177504164aab7fad527b35e0d
[70] WEICHENHAIN, U., KAUFMANN, M., BENZ, A., GOMEZ, G.M., Hydrogen Valleys — Insights into the Emerging Hydrogen Economies Around the World, Fuel Cells and Hydrogen 2 Joint Undertaking (FCH 2 JU), Brussels (2021), https://www.clean-hydrogen.europa.eu/media/publications/hydrogen-valleys-insights-emerging-hydrogen-economies-around-world_en
[71] BHATTACHARYYA, R., SINGH, K.K., GROVER, R.B., BHANJA, K., Nuclear hydrogen production for industrial decarbonization: Creating the business case for the near term, Int. J. Energy Res. **46** 2 (2021) 6929–6943, https://doi.org/10.1002/er.7572
[72] KULMALA, V., The Mankala model is a cornerstone of Finnish energy production, press release, TVO, Helsinki (2023), https://www.tvo.fi/en/index/news/pressreleasesstockexchangereleases/2023/themankalamodelisacornerstoneoffinnishenergyproduction.html
[73] IALENTI, V., Mankala chronicles: Nuclear energy financing and cooperative corporate form in Finland, Nucl. Technol. **207** 9 (2021) 1377–1393, https://doi.org/10.1080/00295450.2020.1868890

[74] DEPARTMENT FOR BUSINESS ENERGY AND INDUSTRIAL STRATEGY, Contracts for Difference and Capacity Market Scheme Update 2020, BEIS, London (2020), https://assets.publishing.service.gov.uk/media/5fdb0740d3bf7f40d7444e86/cfd-cm-scheme-update-2020.pdf

[75] DEPARTMENT FOR BUSINESS ENERGY AND INDUSTRIAL STRATEGY, Guidance on Development Costs and the Nuclear Regulated Asset Base model, BEIS, London (2022), https://assets.publishing.service.gov.uk/media/6384ae9ce90e0778a2122668/development-costs-nuclear-rab-model-guidance.pdf

[76] BARINGA, Financing models for nuclear power plants — European Nuclear Power Plant Case Studies, Ministry of Economic Affairs and Climate Policy (2022), https://open.overheid.nl/documenten/ronl-bd63fed5ed0a01178ce57b9febcf74cd088b5b8b/pdf

[77] WORLD BANK, State and Trends of Carbon Pricing Dashboard, online tool, https://carbonpricingdashboard.worldbank.org

[78] INTERNATIONAL ATOMIC ENERGY AGENCY, Climate Change and Nuclear Power 2022 — Securing Clean Energy for Climate Resilience, IAEA, Vienna (2022), https://www.iaea.org/sites/default/files/iaea-ccnp2022-body-web.pdf

[79] STATISTA, Forecast Impact of Carbon Pricing on Hydrogen Production Costs in the European Union in 2030, https://www.statista.com/statistics/1263414/carbon-pricing-impact-on-h2-production

[80] DENYS, P.-J., KETTLEWELL, W.-J., EU: Proposals Related to the New EU Gas Package (2022), https://www.globalcompliancenews.com/2022/01/18/eu-proposals-related-to-the-new-eu-gas-package-03012022/

[81] DULIAN, M., WILSON, A., Recast EU Directive on Hydrogen and Gas Networks, Briefing — EU Legislation in Progress, European Parliamentary Research Service (2024), https://www.europarl.europa.eu/RegData/etudes/BRIE/2022/729303/EPRS_BRI(2022)729303_EN.pdf

[82] PORTER, M.E., How Competitive Forces Shape Strategy, Harvard Business Review, **57** 2 (1979) 137–145.

[83] EUROPEAN COMMISSION, A Hydrogen Strategy for a Climate-neutral Europe, Communication from the Commission to the European Parliament, the Council, the European Economic and Social Committee and the Committee of the Regions, COM/2020/301 final, Brussels (2020), https://eur-lex.europa.eu/legal-content/EN/TXT/?uri=CELEX:52020DC0301

[84] EUROPEAN COMMISSION, State Aid: Commission approves up to €5.2 billion of public support by thirteen Member States for the second Important Project of Common European Interest in the hydrogen value chain, press release, Brussels (2022), https://ec.europa.eu/commission/presscorner/detail/en/ip_22_5676,

[85] Infrastructure Investment and Jobs Act, H.R.3684, 117th Cong., USA (2021), https://www.congress.gov/bill/117th-congress/house-bill/3684

[86] ENVIRONMENT+ENERGY LEADER, DOE Hydrogen Hub Program Launches (2022),
https://www.environmentenergyleader.com/stories/doe-hydrogen-hub-program-launches,3143
[87] DEPARTMENT FOR BUSINESS ENERGY AND INDUSTRIAL STRATEGY, £102 million government backing for nuclear and hydrogen innovation in the UK, press release (2022),
https://www.gov.uk/government/news/102-million-government-backing-for-nuclear-and-hydrogen-innovation-in-the-uk
[88] UNITED STATES DEPARTMENT OF ENERGY, U.S. National Clean Hydrogen Strategy and Roadmap (2023),
https://www.hydrogen.energy.gov/docs/hydrogenprogramlibraries/pdfs/us-national-clean-hydrogen-strategy-roadmap.pdf
[89] CLARK, K., These Four U.S. Nuclear Plants will Start Producing Clean Hydrogen (2022),
https://www.renewableenergyworld.com/hydrogen/these-four-u-s-nuclear-plants-will-start-producing-clean-hydrogen/
[90] UNITED STATES DEPARTMENT OF ENERGY, DOE Announces $20 Million to Produce Clean Hydrogen From Nuclear Power (2021),
https://www.energy.gov/articles/doe-announces-20-million-produce-clean-hydrogen-nuclear-power
[91] EDF ENERGY R&D UK CENTRE, EDF ENERGY NUCLEAR GENERATION LIMITED, LANCASTER UNIVERSITY, HYNAMICS, ATKINS, EIFER, H2H – Hydrogen to Heysham Feasibility Report, Department of Business, Energy and Industrial Strategy (2019),
https://assets.publishing.service.gov.uk/media/5e4ab9beed915d4ff62d1d55/Phase_1_-_EDF_-_Hydrogen_to_Heysham.pdf
[92] THE WHITE HOUSE, Inflation Reduction Act Guidebook (2023),
https://bidenwhitehouse.archives.gov/cleanenergy/inflation-reduction-act-guidebook/
[93] UNITED STATES DEPARTMENT OF ENERGY, Assessing Lifecycle Greenhouse Gas Emissions Associated with Electricity Use for the Section 45V Clean Hydrogen Production Tax Credit (2023),
https://www.energy.gov/sites/default/files/2023-12/Assessing_Lifecycle_Greenhouse_Gas_Emissions_Associated_with_Electricity_Use_for_the_Section_45V_Clean_Hydrogen_Production_Tax_Credit.pdf
[94] KRUPNICK, A., BERGMAN, A., Incentives for Clean Hydrogen Production in the Inflation Reduction Act, Report 22–13, Resources for the Future, Washington, DC (2022),
https://www.rff.org/publications/reports/incentives-for-clean-hydrogen-production-in-the-inflation-reduction-act/

[95] EUROPEAN COMMISSION, A Green Deal Industrial Plan for the Net-Zero Age, Communication from the Commission to the European Parliament, the European Council, the Council, the European Economic and Social Committee and the Committee of the Regions, COM(2023) 62 final, Brussels (2023), https://eur-lex.europa.eu/legal-content/EN/TXT/?uri=CELEX:52023DC0062

[96] FUEL CELLS AND HYDROGEN 2 JOINT UNDERTAKING, WEEDA, M., Towards a Dual Hydrogen Certification System for Guarantees of Origin and for the Certification of Renewable Hydrogen in Transport and for Heating & Cooling, Final Report of Phase 2, Publications Office of the European Union (2019), https://data.europa.eu/doi/10.2843/46282

[97] IRENA COALITION FOR ACTION, Decarbonising End-use Sectors: Green Hydrogen Certification, IRENA, Abu Dhabi (2022), https://www.irena.org/publications/2022/Mar/The-Green-Hydrogen-Certification-Brief

[98] EUROPEAN COMMISSION, Carbon Border Adjustment Mechanism, https://taxation-customs.ec.europa.eu/carbon-border-adjustment-mechanism_en

[99] UNITED NATIONS ECONOMIC COMMISSION FOR EUROPE, Carbon Neutrality in the UNECE Region — Integrated Life-cycle Assessment of Electricity Sources, United Nations, Geneva (2022), https://unece.org/sites/default/files/2022-04/LCA_3_FINAL%20March%202022.pdf

[100] INTERNATIONAL ATOMIC ENERGY AGENCY, Hydrogen Production Using Nuclear Energy, IAEA Nuclear Energy Series No. NP-T-4.2, IAEA, Vienna (2012).

[101] INTERNATIONAL ENERGY AGENCY, Hydrogen Patents for a Clean Energy Future, IEA, Paris (2023), https://www.iea.org/reports/hydrogen-patents-for-a-clean-energy-future

[102] EL-SHAFIE, M., Hydrogen production by water electrolysis technologies: A review, Results Eng. **20** (2023), https://doi.org/10.1016/j.rineng.2023.101426

[103] BOARDMAN, R.D., WESTOVER, T.L., REMER, S.J., Plan for Scaling Up Hydrogen Production with Nuclear Power Plants. Light Water Reactor Sustainability Program, INL/RPT-22-68155, Idaho National Laboratory (2022).

[104] REVANKAR, S.T., "Nuclear hydrogen production", Storage and Hybridization of Nuclear Energy: Techno-economic Integration of Renewable and Nuclear Energy (BINDRA, H., REVANKAR, S., eds), Academic Press, Cambridge, MA (2019) Ch. 4, https://doi.org/10.1016/C2017-0-00346-4

[105] WENDT, D.S., KNIGHTON, T., BOARDMAN, R., High-Temperature Steam Electrolysis Process Performance and Cost Estimates, INL/RPT-22-66117, Idaho National Laboratory (2022), https://doi.org/10.2172/1867883

[106] HYDROGEN EUROPE, Hydrogen Production and Water Consumption, Hydrogen Europe, Brussels (2020), https://hydrogeneurope.eu/wp-content/uploads/2022/02/Hydrogen-production-water-consumption_fin.pdf

[107] INTERNATIONAL RENEWABLE ENERGY AGENCY, BLUERISK, Water for Hydrogen Production, IRENA, Abu Dhabi (2023), https://www.irena.org/Publications/2023/Dec/Water-for-hydrogen-production

[108] ZHANG, L., WANG, Z., QIU, J., Energy-saving hydrogen production by seawater electrolysis coupling sulfion degradation, Adv. Mater. **34** 16 (2022), https://doi.org/10.1002/adma.202109321

[109] KATEBAH, M., AL-RAWASHDEH, M.M., LINKE, P., Analysis of hydrogen production costs in steam-methane reforming considering integration with electrolysis and CO2 capture, Cleaner Eng. Technol. **10** (2022), https://doi.org/10.1016/j.clet.2022.100552

[110] HIROKI, N., et al., Hydrogen production using thermochemical water-splitting iodine–sulfur process test facility made of industrial structural materials: Engineering solutions to prevent iodine precipitation, Int. J. Hydrogen Energy **46** 43 (2021) 22328–22343, https://doi.org/10.1016/j.ijhydene.2021.02.071

[111] NUCLEAR ENERGY AGENCY, Beyond Electricity: The Economics of Nuclear Cogeneration, OECD Publishing, Paris (2022), https://www.oecd-nea.org/upload/docs/application/pdf/2022-07/7363_cogen.pdf

[112] ISHAQ, H., DINCER, I., A comparative evaluation of three CuCl cycles for hydrogen production, Int. J. Hydrogen Energy **44** 16 (2019) 7958–7968, https://doi.org/7910.1016/j.ijhydene.2019.7901.7249

[113] NATERER, G.F., et al., Progress in thermochemical hydrogen production with the copper–chlorine cycle, Int. J. Hydrogen Energy **40** 19 (2015) 6283–6295, https://doi.org/10.1016/j.ijhydene.2015.02.124

[114] NATERER, G.F., et al., Advances in unit operations and materials for the CuCl cycle of hydrogen production, Int. J. Hydrogen Energy **42** 24 (2017) 15708–15723, https://doi.org/10.1016/j.ijhydene.2017.03.133

[115] LI, H., et al., Canadian advances in the copper–chlorine thermochemical cycle for clean hydrogen production: A focus on electrolysis, Int. J. Hydrogen Energy **45** 58 (2020) 33037–33046, https://doi.org/10.1016/j.ijhydene.2020.09.065

[116] THOMAS, D., BAVEJA, N.A., SHENOY, K.T., JOSHI, J.B., Experimental study on the mechanism and kinetics of $CuCl_2$ hydrolysis reaction of the Cu–Cl thermochemical cycle in a fluidized bed reactor, Ind. Eng. Chem. Res. **59** 26 (2020) 12028–12037, https://doi.org/10.1021/acs.iecr.0c01807

[117] SAYYAADI, H., A conceptual design of a dual hydrogen-power generation plant based on the integration of the gas-turbine cycle and copper chlorine thermochemical plant, Int. J. Hydrogen Energy **42** 48 (2017) 28690–28709, https://doi.org/10.1016/j.ijhydene.2017.09.070

[118] OZBILEN, A., DINCER, I., ROSEN, M.A., A comparative life cycle analysis of hydrogen production via thermochemical water splitting using a Cu–Cl cycle, Int. J. Hydrogen Energy **36** 17 (2011) 11321–11327, https://doi.org/10.1016/j.ijhydene.2010.12.035

[119] AL-ZAREER, M., DINCER, I., ROSEN, M.A., Development and assessment of a novel integrated nuclear plant for electricity and hydrogen production, Energy Convers. Manage. **134** (2017) 221–234,
https://doi.org/10.1016/j.enconman.2016.12.004

[120] BICER, Y., DINCER, I., Life cycle assessment of nuclear-based hydrogen and ammonia production options: A comparative evaluation, Int. J. Hydrogen Energy **42** 33 (2017) 21559–21570,
https://doi.org/10.1016/j.ijhydene.2017.02.002

[121] SCHEIBLEHNER, D., ANTREKOWITSCH, H., NEUSCHITZER, D., WIBNER, S., SPRUNG, A., Hydrogen production by methane pyrolysis in molten Cu–Ni–Sn alloys, Metals **13** 7 (2023) 1310,
https://doi.org/10.3390/met13071310

[122] ABÁNADES, A., RUBBIA, C., SALMIERI, D., Technological challenges for industrial development of hydrogen production based on methane cracking, Energy **46** 1 (2012) 359–363,
https://doi.org/10.1016/j.energy.2012.08.015

[123] McFARLAND, E., PALMER, C., ZENG, J., "Methane pyrolysis for CO2-free hydrogen production", Methane Conversion Routes: Status and Prospects (GALVITA, V., BOS, R., Eds), Vol. 76, Royal Society of Chemistry (2023) Ch. 6,
https://doi.org/10.1039/9781839160257-00148

[124] WNUKOWSKI, M., Methane pyrolysis with the use of plasma: review of plasma reactors and process products, Energies **16** 18 (2023) 6441,
https://doi.org/10.3390/en16186441

[125] BRUCE, S., et al., National Hydrogen Roadmap: Pathways to an Economically Sustainable Hydrogen Industry in Australia, Commonwealth Scientific and Industrial Research Organisation, Canberra (2018).

[126] NUCLEAR REGULATORY COMMISSION, Next Generation Nuclear Plant (NGNP),
https://www.nrc.gov/reactors/new-reactors/advanced/who-were-working-with/licensing-activities/ngnp.html

[127] WEICHENHAIN, U., Hydrogen transportation is key to unlocking the clean hydrogen economy, Roland Berger GmbH, Munich (2021),
https://www.rolandberger.com/en/Insights/Publications/Transporting-the-fuel-of-the-future.html#download

[128] GUIDEHOUSE ENERGY GERMANY GMBH, Covering Germany's Green Hydrogen Demand: Transport Options for Enabling Imports, Guidehouse, Berlin (2022),
https://guidehouse.com/-/media/www/site/insights/energy/2022/transport-options-for-covering-germanys-green-hydrogen-demand.pdf

[129] BRUCEPOWER, Bruce Power to Explore Feasibility of Using Excess Energy for Hydrogen Production (2022),
https://www.brucepower.com/2022/04/07/bruce-power-to-explore-feasibility-of-using-excess-energy-for-hydrogen-production/

[130] NPHYCO, Nuclear Powered Hydrogen Cogeneration — Scenario Definition, D1.2, Rev. A, (2023),
https://www.nphyco.org/uploads/1/4/3/5/143578337/nphyco_d1.2_scenariodefinition.pdf

[131] WORLD NUCLEAR NEWS, OKG Signs Hydrogen Supply Contract (2022),
https://world-nuclear-news.org/Articles/OKG-signs-hydrogen-supply-contract

[132] WORLD NUCLEAR NEWS, MHI to Lead Development of Japanese HTGR (2023),
https://world-nuclear-news.org/Articles/MHI-to-lead-development-of-Japanese-HTGR

[133] MINISTRY OF ECONOMY, TRADE AND INDUSTRY, Overview of Japan's Green Growth Strategy Through Achieving Carbon Neutrality in 2050 (2021),
https://www.mofa.go.jp/files/100153688.pdf

[134] HYDROGENWIRE, Nuclear Power Plant Kola NPP Plans to Produce up to 150 tonnes of Pure Hydrogen per Year (2024),
https://hydrogenwire.com/2024/01/13/nuclear-power-plants-kola-npp-plans-to-produce-up-to-150-tonnes-of-pure-hydrogen-per-year/

[135] ROSATOM, SMR Nuclear Power Plant to be Built in Yakutia by 2028 (2020),
https://fnpp.info/latest-news/smr-nuclear-power-plant-to-be-built-in-yakutia-by-2028

[136] KUCKRO, R., DOE Touts Nuclear-powered Hydrogen Production Projects with Xcel, Constellation, 4 Other Partners (2022),
https://www.utilitydive.com/news/doe-nuclear-powered-hydrogen-production-xcel-constellation-bloom-energy-energy-harbor-aps/636262/

[137] UNITED STATES DEPARTMENT OF ENERGY, H2@Scale, Hydrogen and Fuel Cell Technologies Office,
https://www.energy.gov/eere/fuelcells/h2scale

[138] UNITED STATES DEPARTMENT OF ENERGY, Hydrogen Production Related Links, Hydrogen and Fuel Cell Technologies Office,
https://www.energy.gov/eere/fuelcells/hydrogen-production-related-links

[139] UNITED STATES DEPARTMENT OF ENERGY, Annual Merit Review and Peer Evaluation, Hydrogen Program,
https://www.hydrogen.energy.gov/library/annual-review

[140] WORLD NUCLEAR NEWS, Nine Mile Point Starts Supplying Hydrogen (2023),
https://www.world-nuclear-news.org/Articles/Nine-Mile-Point-starts-supplying-hydrogen

[141] ENERGY HARBOR, Energy Harbor and the Department of Energy Advance Zero-Carbon Hydrogen Production Pilot at Davis-Besse Power Station, news release (2021),
https://www.prnewswire.com/news-releases/energy-harbor-and-the-department-of-energy-advance-zero-carbon-hydrogen-production-pilot-at-davis-besse-power-station-301403337.html

[142] INTERNATIONAL ATOMIC ENERGY AGENCY, Mitigation of Hydrogen Hazards in Severe Accidents in Nuclear Power Plants, IAEA-TECDOC-1661, IAEA, Vienna (2011).

[143] NUCLEAR ENERGY AGENCY, Status Report on Hydrogen Management and Related Computer Codes, Rep. NEA/CSNI/R(2014)8, OECD Publishing, Paris (2014).

[144] HYDROGEN TOOLS, Hydrogen Compared to Other Fuels, https://h2tools.org/bestpractices/gaseous-gh2-and-liquid-h2-fueling-stations/hydrogen-compared-to-other-fuels
[145] CIRRONE, D., MAKAROV, D., PROUST, C., MOLKOV, V., Minimum ignition energy of hydrogen-air mixtures at ambient and cryogenic temperatures, Int. J. Hydrogen Energy **48** 43 (2023) 16530–16544, https://doi.org/10.1016/j.ijhydene.2023.01.115
[146] GIACOMAZZI, E., et al., Hydrogen combustion: Features and barriers to its exploitation in the energy transition, Energies **16** 20 (2023) 7174, https://doi.org/10.3390/en16207174
[147] BREITUNG, W., Analysis methodology for hydrogen behaviour in accident scenarios, Institute for Nuclear and Energy Technologies, Forschungszentrum Karlsruhe, https://h2tools.org/sites/default/files/2019-08/120009.pdf
[148] CHEIKHRAVAT, H., et al., Effects of water sprays on flame propagation in hydrogen/air/steam mixtures, Proc. Combust. Inst. **35** 3 (2015) 2715–2722, https://doi.org/10.1016/j.proci.2014.05.102
[149] GROSSEUVRES, R., COMANDINI, A., BENTAIB, A., CHAUMEIX, N., Combustion properties of H2/N2/O2/steam mixtures, Proc. Combust. Inst. **37** 2 (2019) 1537–1546, https://doi.org/10.1016/j.proci.2018.06.082
[150] VERFONDERN, K., et al., Handbook of Hydrogen Safety: Chapter on LH2 Safety, Fuel Cells and Hydrogen Joint Undertaking (2021), https://hysafe.info/wp-content/uploads/sites/3/2021/04/D39_2021-01-PRESLHY_ChapterLH2-v3.pdf
[151] KUZNETSOV, M., CZERNIAK, M., GRUNE, J., JORDAN, T., Effect of temperature on laminar flame velocity for hydrogen–air mixtures at reduced pressures, paper presented at 5th Int. Conf. Hydrogen Safety (ICHS-5), Brussels (2013).
SCHROEDER, V., HOLTAPPELS, K., Explosion characteristics of hydrogen–air and hydrogen–oxygen mixtures at elevated pressures, paper presented at Int. Conf. Hydrogen Safety (ICHS), Pisa (2005), https://conference.ing.unipi.it/ichs2005/Papers/120001.pdf
[152] ZABETAKIS, M.G., Safety with Cryogenic Fluids, Plenum Press, New York (1967).
[153] EICHERT, H., Gefährdungspotential bei einem verstärkten Wasserstoffeinsatz, Study for the Büro für Technikfolgenabschätzung des Deutschen Bundestags, Deutsches Zentrum für Luft- und Raumfahrt (DLR), Stuttgart (1992).
[154] INTERNATIONAL ATOMIC ENERGY AGENCY, Hydrogen Phenomena During Severe Accidents in Water Cooled Reactors, Training Course Series No. 72, IAEA, Vienna (2021).
[155] EUROPEAN COMMISSION, Equipment for Potentially Explosive Atmospheres (ATEX), https://single-market-economy.ec.europa.eu/sectors/mechanical-engineering/equipment-potentially-explosive-atmospheres-atex_en
[156] EUROPEAN AGENCY FOR SAFETY AND HEALTH AT WORK, Directive 99/92/EC — Risks from Explosive Atmospheres (2018), https://osha.europa.eu/en/legislation/directives/21

[157] EUROPEAN PARLIAMENT AND COUNCIL OF THE EUROPEAN COMMISSION, Directive 2014/34/EU on the Harmonisation of the Laws of the Member States Relating to Equipment and Protective Systems Intended for Use in Potentially Explosive Atmospheres (recast) (2014),
https://eur-lex.europa.eu/eli/dir/2014/34/oj/eng

[158] NATIONAL FIRE PROTECTION ASSOCIATION, Recommended Practice for the Classification of Flammable Liquids, Gases, or Vapors and of Hazardous (Classified) Locations for Electrical Installations in Chemical Process Areas, NFPA 497 (2024),
https://www.nfpa.org/codes-and-standards/nfpa-497-standard-development/497

[159] CANADIAN NUCLEAR LABORATORIES, Gaps and Considerations for Co-Locating or Coupling Hydrogen Production with a Nuclear Asset, internal report, Rep. 153-122800-REPT-004 revision 0, CNL, Ontario (2022).

[160] High Pressure Gas Safety Act, Act No. 204 of 1951 (Japan),
https://www.japaneselawtranslation.go.jp/en/laws/view/4754/en

[161] CLIFFORD CHANCE, Focus on Hydrogen: Japan's Energy Strategy for Hydrogen and Ammonia (2022),
https://www.cliffordchance.com/content/dam/cliffordchance/briefings/2022/08/focus-on-hydrogen-in-japan.pdf

[162] CMS, Facing the Future of Hydrogen: An International Guide (2021),
https://cms.law/en/media/expert-guides/files-for-expert-guides/the-promise-of-hydrogen-an-international-guide-nov-2021

[163] INTERNATIONAL ORGANIZATION FOR STANDARDIZATION, Business Plan of ISO/TC 197 Hydrogen Technologies, ISO, Geneva (2002),
https://www.iso.org/committee/54560.html

[164] AMERICAN SOCIETY OF MECHANICAL ENGINEERS, 2021 ASME Boiler and Pressure Vessel Code, ASME (2021),
https://www.asme.org/codes-standards/bpvc-standards/bpvc-2021

[165] EUROPEAN AGENCY FOR SAFETY AND HEALTH AT WORK, Directive 2014/68/EU — Pressure Equipment (2023),
https://osha.europa.eu/en/legislation/directive/directive-201468eu-pressure-equipment

[166] EUROPEAN AGENCY FOR SAFETY AND HEALTH AT WORK, Directive 2010/35/EU — Transportable Pressure Equipment (2017),
https://osha.europa.eu/en/legislation/directives/directive-2010-35-eu-transportable-pressure-equipment

[167] UNITED NATIONS ECONOMIC COMMISSION FOR EUROPE, Recommendations on the Transport of Dangerous Goods, Model Regulations Rev. 23, ST/SG/AC.10/1/Rev.23 (2023),
https://unece.org/transport/dangerous-goods/un-model-regulations-rev-23

[168] NATIONAL FIRE PROTECTION ASSOCIATION, Hydrogen Technologies Code, NFPA 2 (2023),
https://www.nfpa.org/codes-and-standards/nfpa-2-standard-development/2

[169] INTERNATIONAL ORGANIZATION FOR STANDARDIZATION, Basic Considerations for the Safety of Hydrogen Systems, ISO/TR 15916:2015, ISO, Geneva (2015).

[170] INTERNATIONAL ELECTROTECHNICAL COMMISSION, Explosive Atmospheres — All Parts, IEC 60079:2024, IEC, Geneva (2024).

[171] INTERNATIONAL ORGANIZATION FOR STANDARDIZATION, Hydrogen Generators Using Fuel Processing Technologies, Part 1: Safety, ISO 16110-1:2007, ISO, Geneva (2007).

[172] INTERNATIONAL ORGANIZATION FOR STANDARDIZATION, Hydrogen Generators Using Water Electrolysis — Industrial, Commercial, and Residential Applications, ISO 22734:2019, ISO, Geneva (2019).

[173] INTERNATIONAL ORGANIZATION FOR STANDARDIZATION, Safety of Pressure Swing Adsorption Systems for Hydrogen Separation and Purification, ISO/TS 19883:2017, ISO, Geneva (2017).

[174] INTERNATIONAL ORGANIZATION FOR STANDARDIZATION, Hydrogen Fuel Quality — Product Specification, ISO 14687:2025, ISO, Geneva (2025).

[175] AMERICAN SOCIETY OF MECHANICAL ENGINEERS, Hydrogen Piping and Pipelines, B31.12–2023, ASME, New York (2024).

[176] INTERNATIONAL ORGANIZATION FOR STANDARDIZATION, Cryogenic Vessels — Transportable Vacuum Insulated Vessels of Not More Than 1 000 Litres Volume, Part 1: Design, Fabrication, Inspection and Tests, ISO 21029-1:2018, ISO, Geneva (2018).

[177] INTERNATIONAL ORGANIZATION FOR STANDARDIZATION, Cryogenic Vessels — Transportable Vacuum Insulated Vessels of Not More Than 1 000 Litres Volume, Part 2: Operational Requirements, ISO 21029-2:2015, ISO, Geneva (2015).

[178] INTERNATIONAL ORGANIZATION FOR STANDARDIZATION, Cryogenic Vessels — Large Transportable Vacuum-insulated Vessels, Part 1: Design, Fabrication, Inspection and Testing, ISO 20421-1:2019, ISO, Geneva (2019).

[179] INTERNATIONAL ORGANIZATION FOR STANDARDIZATION, Cryogenic Vessels — Large Transportable Vacuum-insulated Vessels, Part 2: Operational Requirements, ISO 20421-2:2017, ISO, Geneva (2017).

[180] NATIONAL FIRE PROTECTION ASSOCIATION, Compressed Gases And Cryogenic Fluids Code, NFPA 55 (2023),
https://www.nfpa.org/codes-and-standards/nfpa-55-standard-development/55

[181] EUROPEAN HYDROGEN SAFETY PANEL, FUEL CELLS AND HYDROGEN 2 JOINT UNDERTAKING, Safety Planning and Management in EU Hydrogen and Fuel Cells Projects — Guidance Document (2021),
https://www.clean-hydrogen.europa.eu/system/files/2021-09/Safety_Planning_Implementation_and_Reporting_for_EU_Projects-Final.pdf

[182] INTERNATIONAL ELECTROTECHNICAL COMMISSION, Hazard and Operability Studies (HAZOP Studies) — Application Guide, IEC 61882:2016, IEC, Geneva (2016).

[183] INTERNATIONAL ORGANIZATION FOR STANDARDIZATION, Risk Management — Vocabulary, ISO 31073:2022(en), ISO, Geneva (2022).

[184] INTERNATIONAL ATOMIC ENERGY AGENCY, Safety Assessment for Facilities and Activities, IAEA Safety Standards Series No. GSR Part 4 (Rev. 1), IAEA, Vienna (2016).

[185] INTERNATIONAL ATOMIC ENERGY AGENCY, Deterministic Safety Analysis for Nuclear Power Plants, IAEA Safety Standards Series No. SSG-2 (Rev. 1), IAEA, Vienna (2019).

[186] INTERNATIONAL ATOMIC ENERGY AGENCY, Development and Application of Level 1 Probabilistic Safety Assessment for Nuclear Power Plants, IAEA Safety Standards Series No. SSG-3 (Rev. 1), IAEA, Vienna (2024), https://doi.org/10.61092/iaea.3ezv-lp49

[187] INTERNATIONAL ATOMIC ENERGY AGENCY, Development and Application of Level 2 Probabilistic Safety Assessment for Nuclear Power Plants, IAEA Safety Standards Series No. SSG-4, IAEA, Vienna (2010).

[188] INTERNATIONAL ATOMIC ENERGY AGENCY, Format and Content of the Safety Analysis Report for Nuclear Power Plants, IAEA Safety Standards Series No. SSG-61, IAEA, Vienna (2021).

[189] CANADIAN SMALL MODULAR REACTOR ROADMAP STEERING COMMITTEE, A Call to Action: A Canadian Roadmap for Small Modular Reactors, Ottawa (2018), https://smrroadmap.ca/wp-content/uploads/2018/11/SMRroadmap_EN_nov6_Web-1.pdf

[190] GAO, Q., WANG, L., PENG, W., ZHANG, P., CHEN, S., Safety analysis of leakage in a nuclear hydrogen production system, Int. J. Hydrogen Energy **47** 7 (2022) 4916–4931, https://doi.org/10.1016/j.ijhydene.2021.11.099

[191] INTERNATIONAL ATOMIC ENERGY AGENCY, Safety of Nuclear Power Plants: Design, IAEA Safety Standards Series No. SSR-2/1 (Rev. 1), IAEA, Vienna (2016).

[192] NUCLEAR REGULATORY COMMISSION, White Paper—10 CFR 50.59; the Process, Application to Substantial Modifications to Licensee Facilities, and NRC Staff Assessment of Licensee Implementation, https://www.nrc.gov/docs/ML1306/ML13066A266.pdf

[193] REMER, J., BOARDMAN, R., VEDROS, K., CADOGAN, J., Report on the Creation and Progress of the Hydrogen Regulatory Research Review Group, INL/RPT-22-66844 (Rev. 2), Idaho National Laboratory (2023), https://lwrs.inl.gov/content/uploads/11/2024/08/HydrogenRegulatoryResearchReviewGroup.pdf

[194] INTERNATIONAL ATOMIC ENERGY AGENCY, Developments in the Analysis and Management of Combustible Gases in Severe Accidents in Water Cooled Reactors following the Fukushima Daiichi Accident, IAEA-TECDOC-1939, IAEA, Vienna (2020).

[195] INTERNATIONAL ATOMIC ENERGY AGENCY, Stakeholder Engagement in Nuclear Programmes, IAEA Nuclear Energy Series No. NG-G-5.1, IAEA, Vienna (2021).

[196] FORSCHUNGSSTELLE FÜR ENERGIEWIRTSCHAFT E.V., Studie zur Regionalisierung von PtG-Leistungen fur den Szenariorahmen NEP Gas 2020–2030, FfE, Munich (2019).

[197] HUGH, M.J., YETANO ROCHE, M., BENNETT, S.J., A structured and qualitative systems approach to analysing hydrogen transitions: Key changes and actor mapping, Int. J. Hydrogen Energy **32** 10 (2007) 1314–1323, https://doi.org/10.1016/j.ijhydene.2006.10.011

[198] SEYMOUR, E.H., MURRAY, L., FERNANDES, R., Key challenges to the introduction of hydrogen — European stakeholder views, Int. J. Hydrogen Energy **33** 12 (2008) 3015–3020, https://doi.org/10.1016/j.ijhydene.2008.01.042

[199] MURRAY, M.L., SEYMOUR, E.H., ROGUT, J., ZECHOWSKA, S.W., Stakeholder perceptions towards the transition to a hydrogen economy in Poland, Int. J. Hydrogen Energy **33** 1 (2008) 20–27, https://doi.org/10.1016/j.ijhydene.2007.09.020

[200] ANDREASEN, K.P., SOVACOOL, B.K., Energy sustainability, stakeholder conflicts, and the future of hydrogen in Denmark, Renew. Sustain. Energy Rev. **39** (2014) 891–897, https://doi.org/10.1016/j.rser.2014.07.158

[201] SCHMIDT, A., DONSBACH, W., Acceptance factors of hydrogen and their use by relevant stakeholders and the media, Int. J. Hydrogen Energy **41** 8 (2016) 4509–4520, https://doi.org/10.1016/j.ijhydene.2016.01.058

[202] SCHLUND, D., SCHULTE, S., SPRENGER, T., The who's who of a hydrogen market ramp-up: A stakeholder analysis for Germany, Renew. Sustain. Energy Rev. **154** (2022), https://doi.org/10.1016/j.rser.2021.111810

[203] PHAAL, R., FARRUKH, C.J.P., PROBERT, D.R., Technology roadmapping — A planning framework for evolution and revolution, Technol. Forecast. Soc. Change **71** 1 (2004) 5–26, https://doi.org/10.1016/S0040-1625(03)00072-6

[204] INTERNATIONAL ATOMIC ENERGY AGENCY, Developing Roadmaps to Enhance Nuclear Energy Sustainability: Final Report of the INPRO Collaborative Project ROADMAPS, IAEA Nuclear Energy Series No. NG-T-3.22, IAEA, Vienna (2021).

[205] INTERNATIONAL ATOMIC ENERGY AGENCY, Technology Roadmap for Small Modular Reactor Deployment, IAEA Nuclear Energy Series No. NR-T-1.18, IAEA, Vienna (2021).

[206] INTERNATIONAL ATOMIC ENERGY AGENCY, Milestones in the Development of a National Infrastructure for Nuclear Power, IAEA Nuclear Energy Series No. NG-G-3.1 (Rev. 2), IAEA, Vienna (2024), https://doi.org/10.61092/iaea.zjau-e8cs

[207] FUEL CELLS AND HYDROGEN 2 JOINT UNDERTAKING, Hydrogen Roadmap Europe — A Sustainable Pathway for the European Energy Transition, Publications Office of the European Union (2016),
https://data.europa.eu/doi/10.2843/341510

[208] INTERNATIONAL RENEWABLE ENERGY AGENCY, Green Hydrogen Strategy: A Guide to Design, IRENA, Abu Dhabi (2024),
https://www.irena.org/-/media/Files/IRENA/Agency/Publication/2024/Jul/IRENA_Green_hydrogen_strategy_design_2024.pdf

[209] MINISTRY OF ENERGY, Alberta Hydrogen Roadmap, Government of Alberta (2021),
https://www.alberta.ca/hydrogen-roadmap.aspx

[210] MINISTERIO DE ENERGÍA, Hidrógeno Verde — Un Proyecto País,, Santiago de Chile (2022),
https://energia.gob.cl/sites/default/files/guia_hidrogeno_abril.pdf

[211] VICE MINISTERIO DE MINAS Y ENERGÍAS, Hacia la Ruta del Hidrógeno Verde en Paraguay — Marco Conceptual (Towards the Green Hydrogen Roadmap in Paraguay — Conceptual framework), Asunción (2021),
https://minasyenergia.mopc.gov.py/index.php?option=com_content&view=article&id=2064

[212] INTERNATIONAL ENERGY AGENCY, Technology Roadmap — Hydrogen and Fuel Cells, IEA, Paris (2015),
https://www.iea.org/reports/technology-roadmap-hydrogen-and-fuel-cells

[213] DEPARTMENT FOR INTERNATIONAL TRADE, Hydrogen Net Zero Investment Roadmap — Leading the Way to Net Zero (2024),
https://assets.publishing.service.gov.uk/media/65ddc51dcf7eb10015f57f9b/hydrogen-net-zero-investment-roadmap.pdf

[214] NATURAL RESOURCES CANADA, Hydrogen Strategy for Canada — Seizing the Opportunities for Hydrogen: A Call to Action (2020),
https://natural-resources.canada.ca/sites/www.nrcan.gc.ca/files/environment/hydrogen/NRCan_Hydrogen-Strategy-Canada-na-en-v3.pdf

[215] MINISTRY OF INDUSTRY AND TRADE, The Czech Republic's Hydrogen Strategy (2021),
https://www.mpo.cz/assets/cz/prumysl/strategicke-projekty/2021/9/Hydrogen-Strategy_CZ_2021-09-09.pdf

[216] MINISTÈRE DE LA TRANSITION ÉCOLOGIQUE ET SOLIDAIRE, Plan de Déploiement de l'Hydrogène pour la Transition Énergétique (2018),
https://www.ecologie.gouv.fr/sites/default/files/Plan_deploiement_hydrogene.pdf

[217] FRANCE HYDROGÈNE, National Hydrogen Strategy — Harnessing the Hydrogen Sector's Full Potential in Pursuit of France's Decarbonization and Reindustrialization (2023),
https://s3.production.france-hydrogene.org/uploads/sites/4/2023/10/France-Hydrogene_National-Hydrogen-Strategy_EN.pdf

[218] FUHRMANN, M., France Driving Europe's Nuclear Power Renaissance, Mitsui & Co. Global Strategic Studies Institute Monthly Report, Mitsui & Co. Global Strategic Studies Institute, Tokyo (2023), https://www.mitsui.com/mgssi/en/report/detail/__icsFiles/afieldfile/2023/05/24/2303e_fuhrmann_e_1.pdf

[219] MINISTRY OF INNOVATION AND TECHNOLOGY, Hungary's National Hydrogen Strategy — Strategy for the Introduction of Clean Hydrogen and Hydrogen Technologies to the Domestic Market and for Establishing Background Infrastructure for the Hydrogen Industry (2021), https://cdn.kormany.hu/uploads/document/a/a2/a2b/a2b2b7ed5179b17694659b8f050ba9648e75a0bf.pdf

[220] GOVERNMENT OF JAPAN, Japan's Strategic Roadmap for Hydrogen and Fuel Cells (2019) (in Japanese), https://climate-laws.org/documents/strategic-roadmap-for-hydrogen-and-fuel-cells_e56e?id=2019-strategic-roadmap-for-hydrogen-and-fuel-cells_bde0

[221] INTERNATIONAL AFFAIRS OFFICE, INDUSTRIAL SCIENCE AND TECHNOLOGY POLICY AND ENVIRONMENT BUREAU, MINISTRY OF ECONOMY, TRADE AND INDUSTRY, Japan's Green Growth Strategy, Japan Spotlight, Japan Economic Foundation (2021), https://www.jef.or.jp/journal/pdf/237th_Special_Article_01.pdf

[222] GOVERNMENT OF THE RUSSIAN FEDERATION, Roadmap for Development of Hydrogen Energy (2020) (in Russian), http://static.government.ru/media/files/7b9bstNfV640nCkkAzCRJ9N8k7uhW8mY.pdf

[223] DEPARTMENT FOR ENERGY SECURITY AND NET ZERO, UK Hydrogen Strategy (2021), https://assets.publishing.service.gov.uk/media/64c7e8bad8b1a70011b05e38/UK-Hydrogen-Strategy_web.pdf

[224] CONSTELLATION ENERGY, Constellation Starts Production at Nation's First One Megawatt Demonstration Scale Nuclear-Powered Clean Hydrogen Facility, press release (2023), https://www.constellationenergy.com/newsroom/2023/Constellation-Starts-Production-at-Nations-First-One-Megawatt-Demonstration-Scale-Nuclear-Powered-Clean-Hydrogen-Facility.html

[225] INTERNATIONAL ATOMIC ENERGY AGENCY, Deployment Indicators for Small Modular Reactors — Methodology, Analysis of Key Factors and Case Studies, IAEA-TECDOC-1854, IAEA, Vienna (2018).

[226] NUCLEAR HYDROGEN INITIATIVE, Hydrogen Production from Carbon-free Nuclear Energy: Overview of Current Policies and Recommendations for Government Actions (2022), https://cdn.nuclear-hydrogen.org/wp-content/uploads/2022/07/25201728/NHI_NHProduction_Report_07.25.22.pdf

Annex

EXAMPLES OF NATIONAL ROADMAPS FOR THE COMMERCIAL DEPLOYMENT OF NUCLEAR HYDROGEN PRODUCTION

This Annex provides an overview of national strategies and plans in Canada, India and Japan aimed at advancing the commercial deployment of nuclear hydrogen production. These roadmaps highlight the diverse approaches and commitments of different countries in leveraging nuclear technology to produce clean hydrogen, supporting global efforts towards sustainable energy and decarbonization.

A–1. CANADA

Canada is one of the top ten hydrogen producers in the world. It has abundant access to clean, reliable and affordable electricity, with nearly 82% of its electricity coming from non-greenhouse gas emitting sources, namely renewable energy and nuclear power. Given that an important driver for the use of hydrogen in Canada is the reduction of greenhouse gas emissions that hydrogen can offer, it is important for Canada to focus future hydrogen production on economic, low carbon intensity pathways.

On 16 December 2020, Natural Resources Canada released the Hydrogen Strategy for Canada, working in partnership with industry, academic and non-governmental organizations and governments at all levels to identify opportunities and challenges associated with hydrogen deployment [A–1]. The hydrogen strategy lays out the framework and foundational actions that will cement hydrogen as a key pathway to achieve Canada's goal of net zero emissions by 2050 and position Canada as a global industrial leader in clean fuels, with aspirations to increase hydrogen production to 4 Mt of hydrogen per year by 2030 and up to 20 Mt by 2050. In support of the implementation of the hydrogen strategy, Natural Resources Canada established 16 thematic working groups to foster coordination and collaboration among its stakeholder participants and to advance the policy and technical work required to take actions on the strategy's recommendations [A–2]. Two groups are of relevance for this publication: the Electricity Working Group and the Nuclear Working Group.

The Electricity Working Group works to identify challenges and opportunities for stakeholders in the electricity sector. It conducts deep dive research and analysis on topics, including the identification and assessment of:

- Grid impacts from the integration of hydrogen production and use;
- Production opportunities and the potential contribution of the electricity sector in Canada's production of low carbon hydrogen;
- Barriers, gaps and potential solutions from a regulatory, policy, economic and financial perspective.

The Nuclear Working Group explores opportunities to leverage Canada's nuclear expertise and capabilities to support a future low carbon hydrogen economy. It works to showcase opportunities to produce hydrogen with nuclear energy in Canada, and has established four task forces to advance specific activities as follows:

- Production Opportunities Task Force. This task force explores and evaluates opportunities to produce hydrogen with nuclear energy, including assessing the use of 'off-peak' nuclear energy compared with dedicated production, the use of existing large scale nuclear power plants compared with new small modular reactors (SMRs), and the relative size and scale of potential applications for hydrogen produced at or near a nuclear facility. The work of the task force is used to inform the analysis of the subsequent groups.
- Technology and Infrastructure Task Force. This task force provides a technical review and assessment of pathways for hydrogen production from nuclear energy. This includes conventional water electrolysis, high temperature steam electrolysis, thermochemical processes or the use of nuclear heat to reduce emissions associated with steam methane reforming. The task force also explores technologies that could enable hydrogen production, including specific SMR technologies, and also identifies hydrogen derived product production processes with nuclear energy. This group supports the identification of gaps relating to the use of existing infrastructure or equipment to produce hydrogen with nuclear energy.
- Economics, Finance, Business Models and Policy Task Force. This task force explores the economics of producing hydrogen with nuclear energy, and works to identify possible financing mechanisms, project structures and policies that may accelerate demand for nuclear energy to produce hydrogen. The task force also coordinates and seeks inputs from the other working groups and international efforts exploring the economics of producing hydrogen, including the Electricity Working Group.
- Safety, Regulations, Code and Standards Task Force. This task force evaluates and identifies challenges, gaps and other considerations associated with producing hydrogen with nuclear energy from the perspective of safety, regulations, codes and standards. The task force works to identify safety considerations for activities being undertaken internationally. It

also looks at historical activities in Canada relating to the collocation of a hydrogen production plant with a nuclear power plant. Notably, the federal government committed the Bruce Heavy Water Plant in 1968, which operated continuously from April 1973 until March 1998, to the production of reactor grade heavy water [A–3]. This is a unique example that set a precedence for a large scale industrial hydrogen installation operated with shared services and resources close to a nuclear station. The task force released a Canadian Hydrogen Codes and Standards Roadmap in 2025 [A–4], presenting the best available information to identify and prioritize the most critical gaps in deploying the hydrogen economy, highlighting areas where additional expertise may be necessary.

Furthermore, the federal government has pledged to align efforts on nuclear energy with the hydrogen strategy as part of Canada's SMR Action Plan, released in December 2020 [A–5]. The action plan brought together more than 100 partners and outlined approximately 500 actions taken to advance the development, demonstration and deployment of SMRs for multiple applications in Canada and abroad.

A–2. INDIA

Nuclear power is considered by the Government of India to be one of the low carbon energy sources (alongside solar, wind and hydropower) critical to achieving climate action objectives through economy wide decarbonization. Nuclear power is explicitly mentioned as a low carbon energy source in India's Nationally Determined Contributions[1] and long term low carbon development strategy for climate change mitigation [A–7]. In this context, nuclear hydrogen production using commercially mature water electrolysis technology can be a promising, low carbon route to obtaining bulk quantities of hydrogen for industrial decarbonization. This is strongly relevant to all countries like India that are net energy importers today and that have to make vast quantities of energy available to their growing population, decarbonize their economies and simultaneously reduce their dependence on imported fossil fuels and chemical feedstock such as natural gas. It is even more significant in the context of the target date of 2070 for a 'net zero India', which has given rise to an increased momentum in

[1] Nationally Determined Contributions (NDCs) are climate action plans that countries create to outline their efforts to reduce greenhouse gas emissions and adapt to the impacts of climate change. These plans are a key part of the Paris Agreement, where each country commits to specific targets and actions to help limit global warming to well below 2°C, with efforts to limit it to 1.5°C [A–6].

hydrogen economy initiatives in the country. It is widely recognized that India's hydrogen initiatives have to be integrated with its long term energy planning and low carbon energy transition strategies. Consequently, India is adopting a comprehensive approach to transform these initiatives into a plan of action.

Initiatives relating to nuclear hydrogen production in India date back to 2006, when the first draft of a national hydrogen roadmap was released. The emphasis of this roadmap was research, development and pilot scale demonstrations for all aspects of hydrogen economy, through collaborations between industry and academia. It specifically included hydrogen production by nuclear power, particularly by high temperature processes supported by advanced nuclear reactors [A–8].

The first phase of the National Green Hydrogen Mission with a commercial focus was approved in January 2023 and emphasized low carbon hydrogen and ammonia production for domestic consumption using renewable and biomass resources [A–9]. The petroleum refineries and fertilizer sectors have been given mandates to gradually shift to low carbon hydrogen from their current practice of using steam methane reformers to produce the required hydrogen on the site, using renewable energy, following a phased approach [A–10]. A target production capacity of 5 Mt per year of green hydrogen (i.e. hydrogen produced using renewable energy) to be completed by 2030 has been proposed by the NITI Aayog (the Indian Government's planning and advisory organization) as part of this mission [A–9, A–11], although other industry organizations and consortia have called for greater ambition in domestic hydrogen production.

Officially, there is no mention of nuclear assisted hydrogen production in this mission document as of now, thus there is no government endorsed roadmap created specifically for the deployment of related projects. However, need–gap assessments to understand the long term potential for such projects are being conducted and nuclear driven hydrogen production technology is being developed by the research organizations under the Department of Atomic Energy. For considerations of reliability of supply, the energy sources for hydrogen production need to be diversified in a country.

In keeping with these developments and policy articulations, India has been carrying out R&D activities on various aspects of nuclear hydrogen production, including techno-commercial analysis. According to Ref. [A–12], India's existing, upcoming and planned nuclear reactors can power modular water electrolysers to produce 1.8–4.0 Mt of hydrogen per year at annualized production costs of US $3–8.5 per kg of hydrogen, which is competitive with the current renewable hydrogen costs using the same kind of electrolysers. Nuclear hydrogen can meet 6–15% of the current green hydrogen demand of priority sectors in India. Switching to green hydrogen in these sectors can avoid CO_2 emissions of up to 570 Mt per year, which is about 15% of India's current total greenhouse gas emissions. This will make it possible for the nuclear industry to diversify and support the decarbonization beyond the Indian electricity sector. It also creates a stronger case for further expansion of India's nuclear power capacity, which would help overcome the current 85% green hydrogen supply deficit, a percentage that would further increase when newer end use sectors for hydrogen open up.

At present, R&D initiatives in India in the field of nuclear hydrogen production are based on several technologies. This includes alkaline and pure water electrolysis and steam electrolysis process development; intermediate temperature (~550°C) thermochemical cycles (e.g. copper–chlorine cycle); and high temperature (~850–950°C) thermochemical cycles (e.g. iodine–sulphur process and its variants). Advanced nuclear reactor technologies (e.g. molten salt reactors, high temperature reactors, fast breeders) to support efficient electricity generation and high temperature hydrogen production are also being developed domestically, particularly with a view towards effective utilization of India's large domestic thorium reserves [A–13].

The Bhabha Atomic Research Centre has domestically developed highly compact, skid mounted, modular alkaline water electrolysers using advanced porous nickel electrodes in a zero-gap filter-press type construction, capable of producing up to 10 Nm^3/h of high purity (99.9%) hydrogen from a solution of 30% potassium hydroxide dissolved in demineralized water, working at current densities of up to 5000 A/m^2 [A–14]. Owing to its modular construction, plant capacity can be scaled up by adding more electrolysis cells in series and stacks in parallel. The technology has been transferred to private and public sector industries in India for scale-up and deployment.

Hydrogen storage technology development is also an active area of research in the nuclear power sector [A–15], as are research, development and deployment of technologies for the prevention and mitigation of hydrogen related fire hazards, particularly those that might arise in the context of severe nuclear accidents in water cooled reactor systems [A–16].

In accordance with the relative maturities of the various hydrogen technologies, for near term implementation in India a programme based on water electrolysers and ancillary technologies coupled with domestic water cooled nuclear reactors is likely to be adopted as the entry point for the nuclear industry into the hydrogen domain. Demonstration projects are planned at selected power reactor sites, focused on meeting the on-site requirement for hydrogen (e.g. for turbo-generator cooling or coolant water chemistry control) using small scale water electrolyser plants and initial techno-commercial analyses [A–17]. Vendor development and technology incubation for scale up and commercialization of the developed technologies are in progress.

A–3. JAPAN

For Japan to meet its national commitment to achieve carbon neutrality by 2050, a reduction in the emissions from the steel and chemicals sectors, which account for about 25% of total domestic CO_2 emissions, is essential. To achieve this, a large scale and economical supply of hydrogen is necessary in order to displace the high demand by the sectors for fossil fuel [A–18].

Japan has invested in the development of high temperature gas cooled reactor technology for decades. The construction of the high temperature test reactor (30 MWt and 950°C output temperature) was completed with first criticality attained in 1998 and it has since been operated successfully by the JAEA [A–19]. The high temperature test reactor has reached a technology readiness level from which a demonstration reactor can be developed for commercialization as a means of producing nuclear heat and carbon free hydrogen [A–20]. The JAEA has also developed and operated pilot scale hydrogen production facilities based on steam reforming and iodine–sulphur thermochemical methods [A–21].

In 2021, the Ministry of Economy, Trade and Industry launched a demonstration project for large scale nuclear hydrogen production using high temperature gas cooled reactor technology. The ministry decided to invest a total of JPY 43.1 billion (about US $330 million) in the project until 2030. The project aims to achieve three major milestones below by 2030:

— The JAEA will establish the technology of coupling the high temperature test reactor with a steam reforming hydrogen production plant and demonstrate the feasibility of nuclear hydrogen production using the high temperature nuclear heat source.
— Japan will demonstrate prospects for the technological feasibility of carbon free hydrogen production using one or more methods including the iodine–sulphur thermochemical method, the methane pyrolysis method and high temperature steam electrolysis.
— When the project is completed in 2030, the evaluation technology for hydrogen production will be established such that the design margin of error between the forecasted value and the actual measured value will be limited to ±10%.

The ultimate goal of Japan's nuclear hydrogen commercialization roadmap until 2050 is a stable mass supply of hydrogen for industrial uses in steel making, chemistry and other industries through high temperature heat of above 800°C and clean hydrogen production methods.

REFERENCES TO ANNEX

[A–1] NATURAL RESOURCES CANADA, Hydrogen Strategy for Canada — Seizing the Opportunities for Hydrogen: A Call to Action (2020), https://natural-resources.canada.ca/sites/www.nrcan.gc.ca/files/environment/hydrogen/NRCan_Hydrogen-Strategy-Canada-na-en-v3.pdf

[A–2] NATURAL RESOURCES CANADA, Hydrogen Strategy for Canada: Progress Report (2024), https://natural-resources.canada.ca/energy-sources/clean-fuels/hydrogen-strategy/hydrogen-strategy-canada-progress-report

[A–3] CANADIAN NUCLEAR SAFETY COMMISSION, Comprehensive Study Report — Bruce Heavy Water Plant Decommissioning Project (2003), https://www.globalsecurity.org/wmd/library/news/canada/cs-report_e.pdf

[A–4] NATURAL RESOURCES CANADA, Canadian Hydrogen Codes and Standards Roadmap (2025), https://natural-resources.canada.ca/sites/admin/files/documents/2025-02/Canadian%20Hydrogen%20Codes%20and%20Standards%20Roadmap_Feb.12_EN.pdf

[A–5] Canada's Small Modular Reactor (SMR) Action Plan (2020), https://smractionplan.ca/

[A–6] UNITED NATIONS CLIMATE CHANGE, Nationally Determined Contributions (NDCs), https://unfccc.int/process-and-meetings/the-paris-agreement/nationally-determined-contributions-ndcs

[A–7] MINISTRY OF ENVIRONMENT, FOREST AND CLIMATE CHANGE, India's Long-Term Low-Carbon Development Strategy (2022), https://unfccc.int/sites/default/files/resource/India_LTLEDS.pdf

[A–8] MINISTRY OF NEW AND RENEWABLE ENERGY, National Hydrogen Energy Roadmap — Pathway for Transition to Hydrogen Energy for India, National Hydrogen Energy Board, New Delhi (2006) https://policy.asiapacificenergy.org/sites/default/files/abridged-nherm.pdf

[A–9] MINISTRY OF NEW AND RENEWABLE ENERGY, National Green Hydrogen Mission (2023), https://mnre.gov.in/en/national-green-hydrogen-mission/

[A–10] MINISTRY OF POWER, "Ministry of Power notifies Green Hydrogen/Green Ammonia Policy", Press Information Bureau Delhi (issued 17 Feb. 2022), https://www.pib.gov.in/PressReleasePage.aspx?PRID=1799067

[A–11] RAJ, K., LAKHINA, P., STRANGER, C., Harnessing Green Hydrogen: Opportunities for Deep Decarbonization in India, NITI Aayog and RMI, New Delhi (2022), https://www.niti.gov.in/sites/default/files/2022-06/Harnessing_Green_Hydrogen_V21_DIGITAL_29062022.pdf

[A–12] BHATTACHARYYA, R., SINGH, K.K., BHANJA, K., GROVER R.B., Leveraging nuclear power-to-green hydrogen production potential in India: A country perspective, Int. J. Energy Res. (2022), https://doi.org/10.1002/er.8348

[A–13] BHATTACHARYYA, R., SINGH, K.K., GROVER, R.B., BHANJA, K., Nuclear hydrogen production for industrial decarbonization: Creating the business case for the near term, Int. J. Energy Res. **46** 5 (2021) 6929–6943, https://doi.org/10.1002/er.7572

[A–14] SANDEEP, K.C., et al., Experimental studies and modeling of advanced alkaline water electrolyser with porous nickel electrodes for hydrogen production, Int. J. Hydrogen Energy **42** 17 (2017) 12094–12103, https://doi.org/10.1016/j.ijhydene.2017.03.154

[A–15] BHATTACHARYYA, R., Study of the energy utilisation characteristics of solid state hydrogen storage and recovery systems, Int. J. Chem. Eng. Process. **6** 1 (2020) 51–61.

[A–16] BHATTACHARYYA, R., KAMATH, S., SANDEEP, K.C., BHANJA, K., Quantitative consequence analysis of selected accident scenarios at a refuelling station for hydrogen vehicles, Int. J. Renewable Energy Commer. **5** 2 (2019) 59–82.

[A–17] BHATTACHARYYA, R., KAMATH, S., Parametric analysis and optimization of nuclear assisted electrolytic hydrogen production system for on-site applications, Int. J. Chem. Eng. Process. **6** 1 (2020) 14–26.

[A–18] THE MINISTERIAL COUNCIL ON RENEWABLE ENERGY, HYDROGEN AND RELATED ISSUES, Basic Hydrogen Strategy (2023), https://www.meti.go.jp/shingikai/enecho/shoene_shinene/suiso_seisaku/pdf/20230606_5.pdf

[A–19] JAPAN ATOMIC ENERGY AGENCY, Outline of High Temperature Engineering Test Reactor, https://www.jaea.go.jp/04/o-arai/nhc/en/faq/httr.html

[A–20] WORLD NUCLEAR NEWS, MHI to Lead Development of Japanese HTGR (2023), https://www.world-nuclear-news.org/Articles/MHI-to-lead-development-of-Japanese-HTGR

[A–21] NOGUCHI, H., KAMIJI, Y., et al., Hydrogen production using thermochemical water-splitting iodine-sulfur process test facility made of industrial structural materials: Engineering solutions to prevent iodine precipitation, Int. J. Hydrogen Energy **46** 43 (2021) 22328–22343, https://doi.org/10.1016/j.ijhydene.2021.02.071

ABBREVIATIONS

BWR	boiling water reactor
CAPEX	capital expenditures
CCS	carbon capture and storage
CCUS	carbon capture, utilization and storage
CFR	Code of Federal Regulations
CNL	Canadian Nuclear Laboratories
CRI	commercial readiness index
DRI	direct reduction of iron
ETC	Energy Transition Commission
FIT	feed-in tariffs
FSAR	final safety analysis report
GHG	greenhouse gas
HAZID	hazard identification
HAZOP	hazard and operability analysis
HTE	high temperature electrolysis
HTGR	high temperature gas cooled reactor
HTSE	high temperature steam electrolysis
HTTR	high temperature test reactor
IEA	International Energy Agency
IEC	International Electrotechnical Commission
IPCC	Intergovernmental Panel on Climate Change
IRENA	International Renewable Energy Agency
ISO	International Organization for Standardization
JAEA	Japan Atomic Energy Agency
LCOH	levelized cost of hydrogen
LTE	low temperature electrolysis
O&M	operation and maintenance
PEM	polymer electrolyte membrane
PPA	power purchase agreement
PV	photovoltaics
R&D	research and development
RCSs	regulations, codes and standards
SMR	small modular reactor
SOEC	solid oxide electrolysis cell
SSCs	structures, systems and components
SWIFT	structured what if technique
SWOT	strengths, weaknesses, opportunities, threats
TRL	technology readiness level
UNECE	United Nations Economic Commission for Europe

CONTRIBUTORS TO DRAFTING AND REVIEW

Alm-Lytz, K.M.	International Atomic Energy Agency
Arlia, N.	Westinghouse, United States of America
Batra, C.	LucidCatalyst, United Kingdom
Bentaib, A.	Institut de Radioprotection et de Sûreté Nucléaire, France
Berthelot, L.	International Atomic Energy Agency
Bhattacharyya, R.	Bhabha Atomic Research Centre, India
Bradley, E.	International Atomic Energy Agency
Castillo, I.	Canadian Nuclear Laboratories, Canada
Cometto, M.	International Atomic Energy Agency
Constantin, A.	International Atomic Energy Agency
Dardour, S.	International Atomic Energy Agency
Dei, G.	Politecnico di Milano, Italy
Ganda, F.	International Atomic Energy Agency
Grira, A.	Westinghouse, France
Knighton, L.	Idaho National Laboratory, United States of America
Leblanc, D.	Terrestrial Energy, Canada
Li, H.	Canadian Nuclear Laboratories, Canada
Locatelli, G.	Politecnico di Milano, Italy
Martin, C.	Westinghouse, United States of America
Paravano, A.	Politecnico di Milano, Italy
Poghosyan, S.	International Atomic Energy Agency
Reinecke, E.	Forschungszentrum Jülich, Germany

Ryland, D.	Canadian Nuclear Laboratories, Canada
Skliarenko, V.	International Atomic Energy Agency
Stoeva, N.	International Atomic Energy Agency
Terenzi, M.	Politecnico di Milano, Italy
Tomlin, C.	Electric Power Research Institute, United States of America
Trucco, P.	Politecnico di Milano, Italy
Van Heek, A.	International Atomic Energy Agency
Yan, X.	Japan Atomic Energy Agency, Japan

Technical Meeting

Vienna, Austria: 5–7 April 2022

Consultants Meeting

Vienna, Austria: 20–23 September 2022

Structure of the IAEA Nuclear Energy Series*

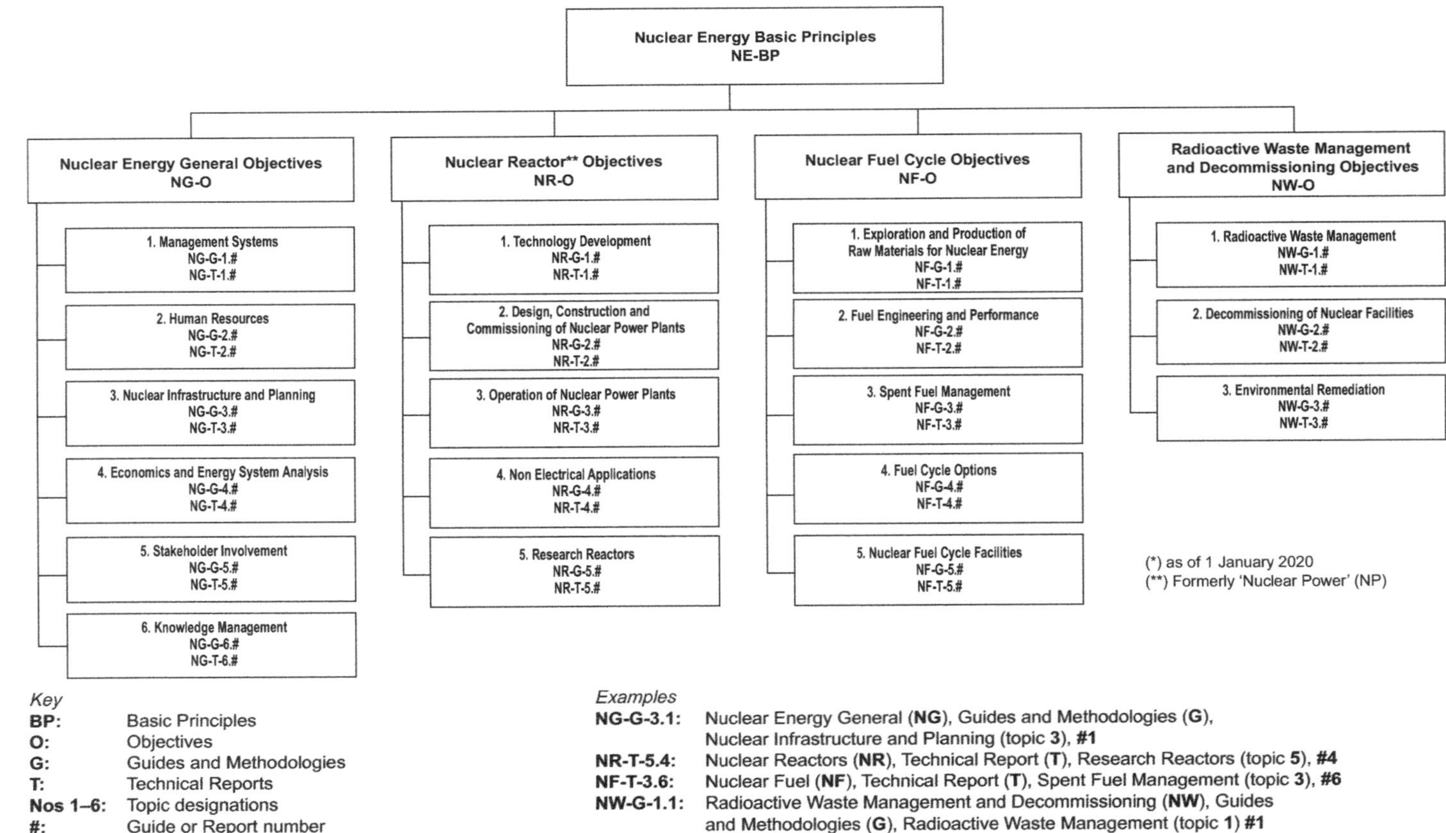

CONTACT IAEA PUBLISHING

Feedback on IAEA publications may be given via the on-line form available at:
www.iaea.org/publications/feedback

This form may also be used to report safety issues or environmental queries concerning IAEA publications.

Alternatively, contact IAEA Publishing:

Publishing Section
International Atomic Energy Agency
Vienna International Centre, PO Box 100, 1400 Vienna, Austria
Telephone: +43 1 2600 22529 or 22530
Email: sales.publications@iaea.org
www.iaea.org/publications

Priced and unpriced IAEA publications may be ordered directly from the IAEA.

ORDERING LOCALLY

Priced IAEA publications may be purchased from regional distributors and from major local booksellers.

Printed and bound by CPI Group (UK) Ltd, Croydon, CR0 4YY
14/07/2026
02167186-0001